Plant Cell Culture

The
Basics

Plant Cell Culture

D. E. Evans

Oxford Brookes University, Oxford, UK

J.O.D. Coleman

Oxford Brookes University, Oxford, UK

and

A. Kearns

Oxford Brookes University, Oxford, UK

Taylor & Francis
Taylor & Francis Group
LONDON AND NEW YORK

© Taylor & Francis, 2003

First published 2003

A CIP catalogue record for this book is available from the British Library.

ISBN 1 85996 320 X

Published by Taylor & Francis
2 Park Square, Milton Park, Abingdon, Oxon, OX14 4RN
270 Madison Ave, New York NY 10016

Transferred to Digital Printing 2007

Typeset by J&L Composition, N Yorks, UK

Publisher's Note
The publisher has gone to great lengths to ensure the quality of this
reprint but points out that some imperfections in the original
may be apparent

Contents

Chapter 8 Protoplast culture   95

Chapter 9 Haploid cultures   121

Chapter 10 Organ and embryo culture   129

Chapter 11 Regeneration of plants and micropropagation

Chapter 12 Somaclonal variation

Chapter 13 Bacterial culture in the plant cell culture laboratory

Abbreviations

2,4-D	2,4-dichlorophenoxyacetic acid
2iP	N-isopentenylaminopurine
ABA	abscisic acid
AFLP	amplified fragment length polymorphism
AGP	arabinogalactan
AP-PCR	arbitrarily primed PCR
BAP	6-benzylaminopurine
BM	basal medium
CaMV	cauliflower mosaic virus
CB	cellulase buffer
CMS	cytoplasmic male sterile/sterility
CUG	3-carboxyumbelliferyl-β-D-galactopyranoside
CWM	Calcofluor White medium
DG1/2	density gradient step 1/2
EB	electroporation buffer
EDTA	ethylenediamine tetraacetic acid
EMS	ethyl methane sulphonate
EPSP	5-enolpyruvylshikimate-3-phosphate
EPSPS	5-enolpyruvylshikimate-3-phosphate synthase
EST	expressed sequence tag
FDA	fluorescein diacetate
GA_3	gibberellic acid
GFP	green fluorescent protein
GM	genetic modification
GUS	β-glucuronidase
HEPA	high-efficiency particulate air
IAA	indole-3-acetic acid
IBA	indole-3-butyric acid
IPAR	6-(γ,γ-dimethylallylamino)purine riboside
lacZ	β-galactosidase
LB	Luria broth
LMP	low melting point
LRR	Leu-rich repeat
LS	Linsmaier and Skoog medium
LUC	luciferase
MAS	marker-assisted selection
MES	2-(N-morpholino)ethanesulphonic acid
MR	rooting medium
MS medium	Murashige and Skoog medium
MSAR	MS medium, Arabidopsis
MUG	4-methylumbelliferyl-β-D-glucuronide
NAA	1-naphthaleneacetic acid
NB	nutrient broth

nptll	neomycin phosphotransferase type II gene
NPTII	neomycin phosphotransferase II
PAR	photosynthetic active radiation
PAT	phosphinothricin acetyltransferase
PCM	protoplast culture medium
PCR	polymerase chain reaction
PCV	packed cell volume
PDA	potato dextrose agar
PEG	polyethylene glycol
PEM	pro-embryogenic cell mass
PEP	phosphoenolpyruvate
PM	protoplasting medium
QTL	quantitative trait loci
RAPD-PCR	random amplified polymorphic DNA-PCR
RFLP	restriction fragment length polymorphism
RLK	receptor-like kinase
RPM	revolutions per minute
SR	sequence repeats
TE	Tris-EDTA buffer
UV	ultraviolet
WB	wash buffer
WM	wash medium
YMA	yeast mannitol agar

Preface

Plant cell and tissue culture is a basic technique of contemporary plant science. Plant cell and tissue culture is used widely in agriculture, horticulture, forestry and biotechnology. Many of the world's staple crops have been bred or propagated using clonal methods based in plant tissue culture. In many parts of the world, products of tissue culture supply the horticulture market. Large-scale programmes are under way to clonally propagate forest trees. Perhaps most importantly, plant genetic modification is only possible using cell and tissue culture techniques. Plant cell and tissue culture is therefore in use globally and routinely by scientists and technicians at every level, in commercial units and in research laboratories. This book is intended to provide the basic techniques and understanding required by students, researchers and technicians.

For many years, plant tissue culture was the domain of highly specialist research scientists and units. Early studies generated numerous methods and a diverse range of media, often with an element of uncertainty – or at least lack of explanation – about why some work when others do not. Texts describing the techniques often reflected that lack of explanation. Here, we have sought to provide the basics of the field to allow newcomers to enter it with understanding. Applications are an equally important 'basic' and we have sought to include description of the range and importance of the technology.

Most of the protocols presented in the book result from our own experience of years of plant tissue and cell culture. Each protocol has been written to include detail and notes that we hope will help the user avoid unnecessary errors and confusion. Inevitably, some protocols are included that involve material or methods we do not use in our own research and in those instances, they are based on the best available protocols. We recognize that any of the protocols may need to be modified to meet the requirements of a given laboratory. Plant material is notoriously variable and the book is intended as a sound basis for workers, not as infallible recipes for every application. Where possible, however, we present variations of protocols that allow the reader to see the options available when designing their own.

The various sections of the book have been written with safety in mind, but users should ensure that they are fully familiar with all the safety requirements of equipment and media. Plant cell and tissue culture is not without risk to the experimenter. Over the years, people have been seriously injured by careless technique in the use of autoclaves or fires started in sterile culture cabinets to name just two examples. Some of the chemicals used are also hazardous. In this respect, the book cannot replace appropriate one-to-one training by a skilled worker; it is intended instead as a manual to accompany such training. In addition, those involved in genetic modification should familiarize themselves with – and abide by – local, national and international rules governing the handling and release of genetically modified plants and micro-organisms. A full risk assessment should always be prepared before work commences. The authors accept no liability whatsoever for any injury or loss resulting from the use of this book or for failures of, or errors in the protocols presented.

We are grateful to Mrs Sheona Bellis for photographing some of our cell and tissue culture experiments as illustrations for the book. We also thank the staff at Bios for their assistance in the publication process from inception to print.

Chapter 1

An introduction to plant cell and tissue culture

1. Plant cell and tissue culture

Plant cell and tissue culture is the cultivation of plant cells, tissues and organs under aseptic conditions in controlled environments. Whilst at one time the techniques of plant cell and tissue culture were the preserve of specialists in the field, culture methods are now well established and widely used in many areas of research and commercial plant science.

The technology of plant cell and tissue culture (sometimes termed *in vitro* culture) is based on a variety of methods that range from the isolation and culture of protoplasts (naked plant cells) to the regeneration and clonal propagation of entire plants. The development of plant and tissue culture as a fundamental science was closely linked with the discovery and characterization of the plant hormones, and has facilitated our understanding of plant growth and development. Furthermore, the ability to grow plant cells and tissues in culture and to control their development forms the basis of many practical applications in agriculture, horticulture and industrial chemistry and is a prerequisite for plant genetic engineering.

1.1 Early experiments in plant tissue culture

It has long been known that many plants show the ability to regenerate tissues and organs and to reproduce vegetatively. Over the centuries, horticulturists have exploited this remarkable regenerative ability by using cuttings to propagate new plants with desirable traits. The date of the first successful plant propagation in horticulture is not known, but by the end of the 19th century, such techniques had become quite sophisticated. In contrast, the birth of plant tissue culture can be accurately dated. The first experiments were undertaken in Graz, Austria, by Gottlieb Haberlandt between 1898 and 1902 and preceded the beginnings of animal cell culture by about 10 years. Haberlandt succeeded in maintaining isolated leaf cells alive for extended periods but the cells failed to divide because the simple nutrient media he used lacked the necessary plant hormones. In the early part of the 20th century, progress in growing excised plant tissues in culture continued with the development of sterile working methods and the discovery of the need for B vitamins and auxins for tissue growth. The first successful experiments to maintain growth and cell division in plant cell culture were those of White (1934), who established isolated tomato roots in aseptic culture. White's medium was simple, containing only sucrose, minerals and a yeast extract, which supplied vitamins. It became clear that under these culture conditions, excised organs like roots were capable of synthesizing the hormones necessary to maintain cell division.

1.2   Developments in culture media

Growing interest in plant tissue culture followed White's experiments. It was discovered that growth and cell division of tissues isolated from a variety of plants could be stimulated by additions to the medium. The first such addition was the plant hormone auxin; later coconut milk (the liquid endosperm of the coconut, also known as coconut water) was used. In the late 1930s, Gautheret and Nobecourt were the first to establish reliable protocols for growing masses of undifferentiated plant tissue; first from willow (*Salix*) and later from carrot (*Daucus carota*). These masses of undifferentiated cells, termed callus, could be re-cultured repeatedly on media containing the plant growth substance (hormone) auxin. At the same time, White, who was also studying callus formation, achieved production of a tobacco (*Nicotiana tobacum*) callus that did not require auxin for growth.

1.3   Embryogenesis in culture

In the 1950s, the work of Skoog and his colleagues on the media requirements for tobacco tissue culture led to many important advances. Included were the discovery of the plant growth substance kinetin, a cytokinin, and the development of an important tissue culture medium, the Murashige and Skoog or MS medium (Murashige and Skoog, 1962).

In a classic paper, Skoog and Miller (1957) demonstrated that by manipulating the auxin to cytokinin balance in the medium, they could control the differentiation of roots and shoots from tobacco pith callus. High concentrations of auxin promoted rooting and high concentrations of cytokinin supported shoot formation, whereas at equal concentrations the tissue remained as callus. This process of organ development is known as organogenesis.

Another technical breakthrough came when Steward *et al.* (1958) used nutrient media enriched with coconut milk to regenerate somatic embryos from callus clumps and cell suspensions of carrot. This process was called somatic embryogenesis (embryos generated from somatic cells) and the somatic embryos formed by this process were fully viable and could be grown to mature plants, capable of flowering and setting seed.

The techniques of *in vitro* regeneration of whole plants by organogenesis stepwise via individual organs (e.g. the differentiation of shoot followed by root) and by somatic embryogenesis are important strategies for clonal propagation that have found application in agriculture and horticulture. Furthermore, regeneration is also an essential component of plant genetic engineering, allowing the production of clones of genetically modified plants.

1.4   Early experiments with suspension cultures

The possibility of culturing individual plant cells in liquid media, and thus allowing the application of standard microbiological methods to their study, was one of the objectives of early work on plant cell cultures. However, it was not until the middle of the 1950s that the successful establishment of suspension cultures was achieved in carrot (Nobecourt, 1955). Cells separated from callus into liquid media and kept constantly agitated were shown to grow and divide to form small aggregates of cells suspended in the medium. Later, in the 1960s, Steward *et al.* (1964) and Vasil and Hildebrandt (1965) demonstrated that fully viable somatic embryos could be regenerated from individual single cells grown in culture. The results of these experiments were taken as proof of totipotency in plants – that a single somatic cell can regenerate to form an entire organism. Totipotency is an unusual property of many plant cells and presumably results from the remarkable plasticity of development shown by plants.

2. Industrial-scale plant cell culture

With the development of protocols for the establishment of plant suspension cultures, there was great interest in exploiting *in vitro* cultures grown on a large scale in bioreactors for the production of speciality organic compounds (e.g. pharmaceuticals) that are plant products. This production system for plant compounds was seen as an advantageous alternative to their extraction from whole plants or their chemical synthesis. For a variety of reasons, this technology was less successful than the equivalent fermentation systems for yeast and industrial-scale culture systems for animal cells. Plant cell cultures grow slowly and are easily damaged by pressure effects in the aeration and agitation required to maintain the cultures. In addition, high-value plant compounds are often the products of complex pathways of secondary metabolism that are actively expressed in differentiated organs or tissues and inactive in undifferentiated cells. In a developing field of research, attempts are being made to overcome the differentiation problem by using either organ cultures like hairy roots (Chapter 10) or by using transgenic cell suspensions that constitutively express genes for the critical enzymes. Recently, the production of high-value compounds (e.g. human therapeutic proteins) in transgenic plants growing in the field without the problems of complex culture cycles and the maintenance of aseptic conditions has attracted considerable attention.

3. Plant tissue culture, plant breeding and crop improvement

Plant cells from which the cell walls have been removed are termed protoplasts (Chapter 2). Techniques using enzymatic degradation of cell walls to obtain large numbers of protoplasts were first devised in the 1960s by Cocking and co-workers. In suitable media, protoplasts can synthesize new cell walls, divide to form small cell colonies and ultimately regenerate whole plants. Protoplasts brought into close contact can be induced to fuse by the application of certain agents. The fusion of protoplasts from different species is used to produce so-called somatic hybrids (sometimes termed cybrids). Somatic hybridization circumvents naturally occurring incompatibility mechanisms that prevent the formation of new species through sexual crosses. This technique has been used in plant breeding to hybridize sexually incompatible species. Perhaps the best-known example of protoplast fusion is the somatic hybrid of potato and tomato, the 'pomato' that was first created in the late 1970s (Melchers *et al.*, 1978). Protoplasts have also proved very useful in genetic transformation, particularly for cereals, and can be generated from intact plant material (leaves, roots) or from suspension cultures.

Tissue culture techniques were also explored as a means of improving crop productivity. These included: the identification of beneficial mutations in clonal lines produced by tissue culture (somaclonal variation, see Chapters 3 and 12), and artificial mutagenesis and the production of disease free plants by meristem tip culture (see Chapter 3).

4.   Plant tissue culture and plant genetic engineering

Development of plant tissue culture as a technique in its own right was augmented in the 1980s by the exciting prospect of deliberately and specifically modifying the characteristics of a plant by genetic manipulation. The description of the mechanism of infection of plants by the soil organism *Agrobacterium tumefaciens* (Chapter 4) was followed by its modification to deliver foreign genes. The first demonstrations of stable plant transformation were then made by both American and European scientists (Horsch *et al.*, 1984; Zambryski *et al.*, 1983). Development of standardized methods of plant transformation followed, using *A. tumefaciens* or *Agrobacterium rhizogenes*, or direct methods like microinjection, electroporation or particle bombardment (Chapter 4). Methods were suitable first for herbaceous dicot plants like tobacco (*Nicotiana tabacum*), petunia (*Petunia hybrida*) and tomato (*Lycopersicon esculentum*) (e.g. Horsch *et al.*, 1985), later for woody dicots and lastly, monocots (e.g. Gasser and Fraley, 1989).

It has already been indicated that the techniques of tissue culture are essential tools for plant genetic modification. Subsequent chapters will describe the production of protoplasts, callus and other materials for transformation and their use to generate clonal populations of transformed plants. The techniques of plant tissue culture are therefore no longer limited to the specialist tissue culture laboratory, but are also the everyday tools of the plant molecular biologist. Materials in routine culture now include all the major food and fibre crops as well as many other species of economic importance, including both angiosperm and gymnosperm trees. *Arabidopsis thaliana* (arabidopsis), the subject of the first plant genome project to be completed and a model species in plant science, is also widely used in cell and tissue cultures.

Techniques of plant tissue culture have, therefore, been developed over more than half a century and include many species. Often, published protocols have been presented without a clear rationale for their use, or without attempting to explain why modifications were made for a particular species. This volume seeks to present techniques that work routinely, together with sufficient background detail to permit a new worker to understand and make rational choices in order to modify the protocols for their own use.

References

Gasser, C.S. and Fraley, R.T. (1989) Genetically engineering plants for crop improvement. *Science* **244**: 1293–1299.

Horsch, R.B., Fraley, R.T., Rogers, S.G., Sanders, P.R., Lloyd, A. and Hoffman, N. (1984) Inheritance of functional foreign genes in plants. *Science* **223**: 496–498.

Horsch, R.B., Fry, J.E., Hoffman, N., Eichholtz, D., Rogers, S.G. and Fraley, R.T. (1985) A simple and general method for transferring genes into plants. *Science* **227**: 1229–1231.

Melchers, G., Sacristan, M.D. and Holder, A.A. (1978) Somatic hybrid plants of potato and tomato regenerated from fused protoplasts. *Carlsberg Res. Commun.* **43**: 203–218.

Murashige, T. and Skoog, F. (1962) A revised medium for rapid growth and bioassays with tobacco tissue cultures. *Physiol. Plant* **15**: 473–497.

Nobecourt, P. (1955) *Bull. Soc. Botan. Suisse* **65**: 475–480.

Skoog, F. and Miller, W. (1957) Chemical regulation of growth and organ formation in plant tissues cultured *in vitro*. In: *Biological Action of Growth Substances* (ed. H.K. Porter). Symposium of the Society for Experimental Biology, Cambridge University Press, Cambridge, 11, pp. 118–131.

Steward, F.C., Mapes, M.O. and Mears, K. (1958) Growth and organized development of cultured cells. *Am. J. Bot.* **45**: 705–708.

Steward, F.C., Mapes, M.O., Kent, A.E., Holsten, R.D. (1964) Growth and development of cultured plant cells. *Science* **143**: 20–27.

Vasil, V. and Hildebrandt, A.C. (1965) Differentiation of tobacco plants from a single isolated cell in microculture. *Science* **146**: 76–77.

White, P.R. (1934) Potentially unlimited growth of excised tomato root tips in a liquid medium. *Plant Physiol.* **9**: 585–600.

Zambryski, P., Joos, H., Genetello, C., Leemans, J., Vanmontagu, M., Schell, J. (1983) TI-plasmid vector for the introduction of DNA into plant-cells without alteration of their normal regeneration capacity. *EMBO J.* **2**: 2143–2150.

Chapter 2

Basic plant biology for cell culture

1. Tissues and organs

Before considering the practical aspects of plant tissue culture, it is important to be aware of the way in which plants grow and develop. Their unique growth form, in which discrete areas of cell division remain active throughout the life of the plant, contrasts markedly with animal development. Understanding the zones of cell division and the way in which their activity is initiated and regulated helps to put into context the principles and methods of plant cell and tissue culture.

Plants show a wide diversity of form and vary in the number and types of organs they possess. The major groups of organs, leaves, roots and stems are present in most plants (although all may be very highly modified). Flowers, fruit and storage organs, like tubers, formed at specific stages in the plant's life cycle, also show a variety of forms. Organs may be cultured intact (organ culture) or divided to initiate callus (an undifferentiated cell mass) that can differentiate to form organs and embryos. Sections of leaf, for instance, are often used as the host tissue for genetic transformation and embryos are regenerated from the transgenic tissue.

2. Plant tissues

Plants consist of specialized cell types with differing functions. Plant organs are made of three types of tissue: dermal tissue (the epidermis or outer cell layer), vascular tissue (the transport system of the plant) and ground tissues (all the remaining cells). The type and quantity of each tissue varies for each organ.

2.1 Ground tissues

Ground tissues lie under the epidermis and contribute to structural strength and function. Parenchyma cells are the most abundant cells found throughout the plant and form the bulk of organs like leaves, roots and stems. They have thin flexible cell walls and are initially cuboid, later becoming nearly spherical or cylindrical. In mature tissue, their shape is constrained by surrounding cells. Parenchyma cells contain starch grains and other food reserves and have large vacuoles. Leaf parenchyma cells, termed chlorenchyma, have chloroplasts and are photosynthetic. When the plant is wounded, parenchyma cells divide to form a callus that fills the wound site and may eventually form new tissues to repair the wound. Wound-like callus is frequently formed in plant tissue cultures. Collenchyma cells are similar to parenchyma, but have thickened cell walls and are often found as flexible support beneath the epidermis. Sclerenchyma is a supporting

tissue found in organs that have completed lateral growth. It is made up of
dead cells with even, lignified secondary cell walls and is organized either
as sclereids (small groups of cells) or as fibres (long strands of elongate
cells).

2.2   Dermal tissues

All organs are surrounded by dermal tissue that forms a protective layer
known as the epidermis. It consists of parenchyma or parenchyma-like
cells, and usually forms a complete covering, except where specialized
pores (e.g. for gas exchange) are present. The epidermis protects the tissues
beneath it from mechanical damage and invasion by pathogens. In roots,
the epidermis is involved in nutrient uptake and may possess root hairs.

2.3   Vascular tissue

The vascular tissue made up of xylem and phloem forms the conducting
tissues for fluids through the plant. Xylem carries predominantly water
and dissolved minerals from root to shoot; phloem carries solutes like
sugars and amino acids from sites of synthesis or storage to sites of
storage or use.

2.4   Callus

Sometimes, a wounded plant will produce a proliferation of undifferentiated cells at the wound site. This mass of cells is known as a callus. A similarly disorganized mass of undifferentiated cells produced in tissue culture
is also termed callus. Callus culture is commonly used as the starting point
for various techniques in plant tissue culture, as callus can be induced to
differentiate to form organs and embryos.

3.   Meristems

Meristems are regions of actively dividing cells that give rise to cells that differentiate into new tissues of the plant. *Figure 2.1* shows the location of
meristems in a dicotyledonous angiosperm. Apical meristems occur near the
tips of roots and shoots and are primarily responsible for length extension
of the plant body. This type of growth is called primary growth. Meristems
contain self-perpetuating cells called initials (the equivalent of stem cells in
animals), which retain the ability to divide and generate new cells throughout the life of the meristem. Primary meristems that produce the tissues of
the stem and root originate in the apical meristems. The primary meristems
are the protoderm (that forms the epidermis or outer cell layer of the shoot),
the procambium (that forms phloem and xylem) and the ground meristems
(that form parenchyma).

   Lateral meristems, the vascular cambium and the cork cambium,
produce the secondary tissues and are responsible for increasing the girth of
the plant (secondary growth). The vascular cambium (or cambium) occurs
in stem and root, and in mature stems is extended laterally to form a complete cylinder of cells that forms new phloem and xylem. Cork cambium
arises in the outer layers of the stems of woody plants typically as a cylinder
surrounding the inner tissues. Cells of cork cambium divide to form an
outer layer, cork and an inner layer the secondary cortex or phelloderm.

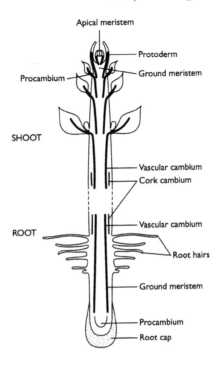

Apical meristem

Protoderm

Ground meristem

Procambium

SHOOT

Vascular cambium

Cork cambium

Vascular cambium

ROOT

Root hairs

Ground meristem

Procambium

Root cap

Figure 2.1

The location of meristems in a dicot. From Lack, A.J. and Evans, D.E. (2001) Instant Notes in Plant Biology. *© BIOS Scientific Publishers Ltd., 2001.*

Cork, cork cambium and the phelloderm together form an outer protective layer known as the periderm.

Meristem culture is used to obtain, maintain and propagate disease-free plants. Viruses, for example, do not invade the meristem and virus-free plants can be produced from cultures of meristematic explants. Meristem culture has been used successfully to eliminate viruses from several species including potato, sugarcane and strawberry.

4. Plant reproductive tissues

Reproductive tissues are of particular interest because they contain haploid cells with the gametic (n) number of chromosomes rather than the ($2n$) found in diploid somatic cells. By the application of cell culture techniques, haploid cells can generate haploid plants (see Chapter 9). Haploid plants can double their chromosome set to form homozygous dihaploid plants. Dihaploids are important elite parents used in breeding programmes for cultivar improvement in several important crops including rice, wheat and *Brassica* species. In angiosperms, the reproductive organ is the flower (*Figure 2.2*). Haploid pollen (the microspore) is borne in stamens, which comprise an anther supported by a filament. Haploid lines are obtained either by culturing excised anthers or by the culture of isolated microspores. Ovules are more complex and contain an embryo sac in which the haploid embryo nucleus is closely associated with diploid cells. Ovule culture has therefore not been favoured as a means to obtain haploid plants.

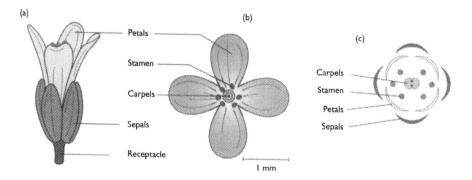

(a)

Petals

Stamen

Carpels

Sepals

Receptacle

(b)

Carpels

Stamen

Petals

Sepals

1 mm

(c)

Figure 2.2

The flower of Arabidopsis thaliana. *From Lack, A.J. and Evans, D.E. (2001) Instant Notes in Plant Biology.* © BIOS Scientific Publishers Ltd., 2001.

In most gymnosperms, microsporangia (anthers) are borne in strobili (cones) on the underside of specialized leaves (*Figure 2.3*). The female reproductive structure is also a cone, much larger than the strobili, with woody scales.

5.   The embryo and embryogenesis

The formation of a plant embryo begins with the zygote, formed from fertilization of an egg cell contained within the embryo sac, by a sperm cell derived from pollen. A second sperm nucleus fertilizes an adjacent cell to form endosperm tissue that will supply nutrients to the developing embryo within the seed.

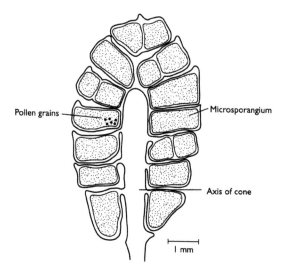

Pollen grains

Microsporangium

Axis of cone

1 mm

Figure 2.3

The strobilus, or male cone of a gymnosperm showing the location of the microspores (pollen). From Lack, A.J. and Evans, D.E. (2001) Instant Notes in Plant Biology. © BIOS Scientific Publishers Ltd., 2001.

The basic plan of the plant is established soon after the ovule is fertilized, in the early stages of the development of the embryo. Embryogenesis in *Arabidopsis thaliana*, a dicot, is shown in *Figure 2.4*. The fertilized ovule divides to give an apical cell and a basal cell. The basal cell forms the suspensor, connecting the embryo to maternal tissue and the root cap meristem. The apical cell undergoes many cell divisions. The first stage is the octant stage (named from the eight cells formed in two tiers). This is followed by the dermatogen stage (where tangential cell divisions have occurred creating tissue layers). Finally the heart-shaped embryo is formed. This contains the origins of all the major structures of the seedling. The lobes of the heart shape are the cotyledons; between them lies the shoot meristem. The centre of the heart forms the hypocotyl and the lower layers form the root.

The process of embryo formation from zygotes is known as zygotic embryogenesis. Embryos formed in culture from somatic tissue pass through morphological stages of development similar to zygotic embryos. This process is known as somatic embryogenesis.

6. Development of tissues

The concentric rings of cells laid down in meristems are initially similar in form, non-vacuolate, isodiametric (roughly cuboid) and with thin cell walls. Subsequently they form all the tissues of the plant. This involves major changes in gene expression. The first stage of this is determination, in which the cell becomes established on a pathway of change, but physical changes are not yet detectable. At this stage, the cell is committed to a pathway of development. The cell then becomes differentiated to its new form and function, losing some characteristics and gaining others. In plant cell cultures, both determination and differentiation are frequently reversible given suitable treatments and pathways of determination may be manipulated to generate new tissues and organs, or to induce embryogenesis.

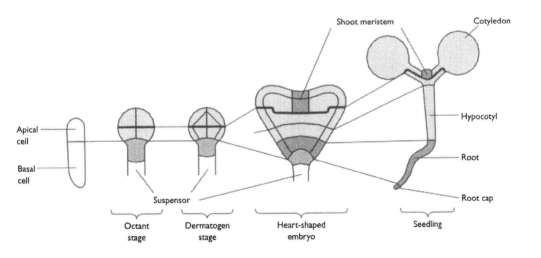

Figure 2.4

Embryogenesis in a typical dicotyledon. Reproduced with permission from Laux, T. and Jurgens, G. (1997) Plant Cell 9: 989–1000. © 1997, American Society of Plant Biologists.

7.   Protoplasts

With very few exceptions, plant cells are enclosed by a cell wall. Protoplasts are single cells that have been stripped of their cell walls. The resulting cell is spherical and consists of the original cell contents surrounded by the plasma membrane. In the absence of a cell wall, protoplasts are vulnerable to osmotic lysis, but they are more amenable than whole cells to a variety of cell and tissue culture techniques, including somatic hybridization and genetic manipulation. Protoplasts can be isolated from a variety of whole-plant tissues and from plant tissue cultures like callus and suspension cultures by enzymatic digestion of cell walls (see Chapter 8).

8.   Media, nutrients and requirements for growth

The culture medium is one of the most important components of plant cell and tissue culture methods. The successful application of plant tissue culture procedures largely depends on a culture medium with the right composition. Media compositions therefore vary with the particular culture technique being employed. For example, procedures like regeneration of whole plants from cells or tissues require media with different compositions to initiate each phase of the developmental sequence. In addition to its composition, another important function of the culture medium is to provide the right physical environment for cells and tissues to grow. Solid media, for example, perform a function like soil by providing a physical support matrix where tissue explants can maintain contact with air for gaseous exchange or where regenerated plantlets can root. On the other hand, growing suspension cultures in agitated liquid culture enables the cells to maintain maximum exposure to the ingredients of the medium and facilitates gaseous exchange.

All living plant cells require water, nutrients (inorganic mineral elements and organic compounds) and plant growth regulators (hormones) for sustained growth and development. The components of plant tissue culture media generally reflect these requirements and several specific formulations (e.g. Murashige and Skoog, MS medium) used routinely for culture have similar basic compositions (see *Table 2.1* and Murashige and Skoog, 1962).

The requirements for plant cell culture may be subdivided into three groups: inorganic (mineral) nutrients, organic nutrients and plant growth regulators.

8.1   Inorganic nutrients

The inorganic nutrients are mineral elements and based on their essential concentrations are usually classified into two groups: macroelements that are present in large supply (mM concentrations) and microelements or trace elements that are essential only at very low concentrations (μM).

Macroelements

Nitrogen (N) Nitrogen is essential for plant life, as it is an integral constituent of many important biological compounds, for example, amino acids, proteins, nucleic acids, chlorophyll and many other biomolecules. In plant culture media, inorganic nitrogen is usually supplied in two forms, in an oxidized form as the nitrate anion (NO_3^-) and in a reduced form as the

Table 2.1. Components of Murashige and Skoog's medium (Murashige and Skoog, 1962)

Component		Molarity in final medium
Major inorganic nutrients (macronutrients)	NH_4NO_3	2.06×10^{-2}
	KNO_3	1.88×10^{-2}
	$CaCl_2.2H_2O$	3.00×10^{-3}
	$MgSO_4.7H_2O$	1.50×10^{-3}
	KH_2PO_4	1.25×10^{-3}
Trace elements (micronutrients)	KI	5.00×10^{-6}
	H_3BO_3	1.00×10^{-4}
	$MnSO_4.4H_2O$	9.99×10^{-5}
	$ZnSO_4.7H_2O$	2.99×10^{-5}
	$CuSO_4.5H_2O$	1.00×10^{-7}
	$CoCl_2.6H_2O$	1.00×10^{-7}
Iron	$FeSO_4.7H_2O$	1.00×10^{-4}
	$Na_2EDTA.2H_2O$	1.00×10^{-4}
Vitamins	Myo-inositol	4.90×10^{-4}
	Nicotinic acid	4.66×10^{-6}
	Pyridoxine-HCl	2.40×10^{-6}
	Thiamine-HCl	3.00×10^{-7}
Organic nitrogen source	Glycine	3.00×10^{-5}
Carbon source	Sucrose	8.80×10^{-2}

ammonium cation (NH_4^+). The concentration of total inorganic nitrogen is usually 20–40 mM. Most cultures seem to grow best when supplied with both inorganic forms. In some media formulations organic nitrogen is also added as amino acids or protein hydrolysates. These organic forms of nitrogen cannot fully substitute for the inorganic forms.

Sulphur (S) Like nitrogen, sulphur is a vital component of proteins in the amino acid residues of cysteine and methionine. Both these amino acids are also precursors for other essential sulphur compounds such as vitamins and co-factors. Sulphur is also present in iron–sulphur proteins, which are important redox mediators of chloroplasts and mitochondria. In plant culture media, sulphur is usually supplied at 1–3 mM as the sulphate anion (SO_4^-) in combination with magnesium as magnesium sulphate.

Phosphorus (P) Phosphorus is a constituent of the nucleic acids DNA and RNA and is abundant in the phospholipids of biomembranes. It is also present in energy-rich phosphate esters like ATP and is an important substrate or product in several metabolic reactions. In plant culture media, phosphorus is supplied as the phosphate anion (PO_4^-) as sodium or potassium dihydrogen phosphate at a concentration of 1–3 mM.

Calcium (Ca) Calcium is an important modulator of enzyme action and is required for cell wall synthesis and stabilization. Calcium is used in plant tissue culture media at concentrations of 1–3 mM and is added either as the chloride or nitrate salt.

Magnesium (Mg) Magnesium acts as the central atom in the chlorophyll molecule, in the aggregation of ribosomal subunits, which is required for protein synthesis and as an essential cofactor for many enzymatic reactions.

Magnesium may also play a role in cation–anion balance. In plant cell and tissue culture media, magnesium is usually added as magnesium sulphate in concentrations of 1–3 mM.

Potassium (K) Potassium is the major cation in plants playing several vital roles. Potassium functions in osmoregulation, cation–anion balance and pH stabilization, as an activator for many enzyme-catalysed reactions and is involved in several key steps in the translation process of protein synthesis. Potassium in tissue culture media may vary with the plant species in culture, but is usually supplied at 20–30 mM in combination with nitrate as potassium nitrate.

Microelements

Iron (Fe) Iron is required for chlorophyll synthesis and functions in redox reactions as a constituent of cytochromes and iron–sulphur proteins. Ferric iron (Fe{III}) is practically insoluble at physiological pHs (the solubility constant for $Fe(OH)_3$ is 10^{-39}) and chelators such as ethylenediamine tetraacetic acid (EDTA) are necessary to keep ferric iron in solution. Fe-EDTA is used in culture media to make iron available over a wide pH range.

Boron (B) Boron is a component part of some structural complexes of cell walls and is required for cell division in the apical meristems of roots. In plant culture media, boron is supplied as boric acid.

Cobalt (Co) It is not clear whether cobalt has any direct function in higher plants. However, cobalt is added to plant culture media at around 0.1 μM.

Copper (Cu) The main function of copper is as protein bound copper in redox reactions. Copper is added to culture media at 0.1 μM as copper sulphate. It is toxic at higher concentrations.

Iodine (I) Iodine is not considered as an essential microelement, but is thought to be a beneficial microelement that improves the growth of roots and callus.

Manganese (Mn) Manganese is found in metalloproteins, where it has a structural role or acts in a redox capacity. Manganese is usually used in culture media at 5–30 μM as manganese sulphate.

Molybdenum (Mo) Molybdenum is a cofactor of a few redox enzymes including nitrate reductase, which is involved in the conversion of nitrate to ammonia. It is usually supplied at 0.1 μM as sodium molybdate.

Zinc (Zn) Zinc is required for the activity of various types of plant enzymes including dehydrogenases and RNA and DNA polymerases. Zinc is added to culture media at 5–30 μM as zinc sulphate.

8.2 Organic nutrients

While green plants are autotrophic (photosynthetic), most culture systems, at least in the early stages, are heterotrophic and require an organic source of carbon and energy. Cultures are often grown with illumination to initiate the development of chloroplasts and photosynthesis. In addition to an organic source of carbon and energy, plant cultures may also require other complex organic molecules for healthy growth.

Sugars

Sugars are added to plant cultures to supply carbon and energy. Sucrose is the most commonly used sugar in plant culture media, but glucose, fructose and sorbitol are used in some formulations. In addition to its metabolic role, sucrose also acts as an osmoticum and together with the inorganic nutrients helps to balance the osmotic potential of the culture medium.

Vitamins and cofactors

Added vitamins are not essential for plant cell and tissue cultures, although vitamin B_1 (thiamine) is considered beneficial for cultures of some species. However, biotin, pantothenic acid, nicotinic acid (niacin), pyridoxine (pyridoxol; vitamin B_6), folic acid, ascorbic acid (vitamin C) and tocopherol (vitamin E) are added to some media. A single addition of yeast extract is sometimes used as a source of vitamins.

The sugar alcohol *myo*-inositol is frequently used in media for monocots, gymnosperms and some dicots. *Myo*-inositol plays a role in cell wall and membrane development.

Complex organic supplements

Complex additions such as banana powder and the liquid endosperm of the coconut (known as coconut milk or coconut water) are frequently added to some media to improve growth. Although coconut water is thought to provide growth regulators and other organic compounds, it is not known which components of this supplement are effective in improving the growth of cultures. Because of this uncertainty, some experts discourage the use of these undefined additions.

8.3 Plant growth regulators

There are five main classes of plant growth regulators or plant hormones that co-ordinate growth and differentiation in plants: auxins, cytokinins, gibberellins, abscisic acid and ethylene. Ethylene is principally involved with abscission, floral senescence and fruit ripening and is rarely used in plant tissue cultures. Of the other four classes of plant hormones, auxins and cytokinins are used frequently in plant tissue cultures and gibberellins and abscisic acid are used occasionally.

Auxins

The major auxin in plants is indole-3-acetic acid (IAA). Auxin is synthesized in the stem and root apices and transported along the plant axis. The primary action of auxin is to stimulate the elongation growth of cells. Shoot growth is stimulated by 10^{-6}–10^{-7} M auxin and root elongation by much lower concentrations (10^{-9}–10^{-10} M).

Auxin also stimulates cell division and differentiation and together with cytokinins regulates several developmental processes. For example, auxin induces lateral root formation in stem cuttings and the differentiation of roots and shoots from callus cultures is controlled by the auxin/cytokinin balance. Similarly, apical dominance, the dominant growth of an apical bud seen in many plants, is also auxin/cytokinin regulated.

Auxin treatment causes the formation of parthenocarpic (seedless) fruits in some species (e.g. strawberry, tomato, cucumber, pumpkin and citrus). Senescence and abscission of mature leaves, fruits and flowers is inhibited by auxin; however, abscission of young fruits is enhanced by auxin treatment.

In plant cell and tissue cultures, auxin together with cytokinin are used to control differentiation and morphogenesis. However, the naturally occurring auxin is seldom used in plant tissue cultures and the synthetic auxins 2,4-dichlorophenoxyacetic acid (2,4-D), 1-naphthaleneacetic acid (NAA), indole-3-butyric acid (IBA) and p-chlorophenoxyacetic acid are the most common alternatives (*Figure 2.5*).

The auxin concentration is a critical factor in plant cell and tissue cultures and the optimum concentration varies from species to species. In most cases, synthetic auxins were first identified from herbicide screens (e.g. 2,4-D is a common herbicide) and at higher concentrations may cause severe growth abnormalities or completely inhibit growth.

Cytokinins

Cytokinins promote growth and development, delay senescence and act with auxins (see above) to control growth and development. The first cytokinin to be identified was kinetin, which was isolated from samples of degrading herring sperm DNA that caused large increases of cell division in tobacco cell cultures. Subsequently, the naturally occurring cytokinin zeatin was isolated from the kernels of corn (*Zea mays*). Chemically cytokinins are N_6-substituted derivatives of the nitrogenous purine base adenine (*Figure 2.5*). The cytokinins most frequently used in tissue culture media include kinetin, benzylaminopurine, zeatin, 6-(γ,γ-dimethylallylamino) purine and adenine.

Gibberellins

In plants, the major action of gibberellins is stimulation of stem elongation and flowering. Gibberellins are also involved in the mobilization of reserves from the endosperm in the initial stages of embryo growth and germination of cereal seeds. Gibberellins are a very large chemical family with over 90 different gibberellins recognized in plants. Each gibberellin is distinguished by a numerical subscript, for example GA_3. In plant tissue cultures, gibberellins are used to induce organogenesis especially adventitious root formation. Only gibberellic acid GA_3, GA_4 and GA_7 are commonly used in plant cultures.

Abscisic acid

In plants, abscisic acid (ABA) is a sesquiterpenoid (15 carbons) primarily involved in water stress responses, induction of storage protein synthesis in seeds and in seed germination. In plant tissue cultures, ABA is used to enhance somatic embryogenesis.

8.4 Support matrices

In some plant cell and tissue culture systems, a solid or semi-solid matrix is needed to support tissue explants while allowing contact with the medium. Although a liquid medium is useful for many culture systems, a frequently encountered problem is the condition called hyperhydricity. Hyperhydricity is a physiological and morphological disorder of plant tissue cultures, where

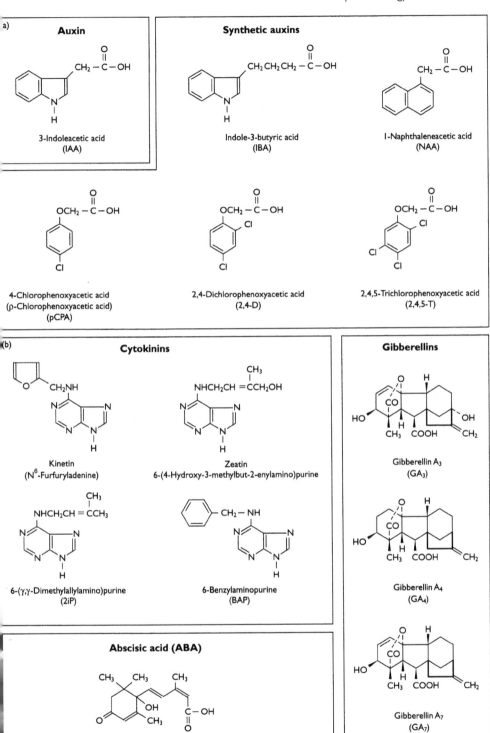

Figure 2.5

The principal plant growth regulators used in tissue culture media. (a) Auxin (IAA) and synthetic auxins, (b) cytokinins, gibberellins and abscisic acid.

plants or tissue grow poorly, have a high water content and a glassy or brittle appearance. A support matrix may be created from materials such as filter paper, rafts of polypropylene or discs of polyurethane foam. However, the most commonly used support matrix is formed from gelling agents such as agar, agarose, gelatine or gellan gums. Some gelling agents are also known to induce hyperhydricity.

Agar

Agar is the most commonly used gel in plant tissue culture. It is a complex polysaccharide extracted from red algae (*Rhodophyceae*) and comprises two fractions, agarose (70%) and agaropectin (30%). Agarose is the gelling fraction and consists of a polymer of alternating D-galactose and 3,6-anhydrogalactose monosaccharide units. Agaropectin is the non-gelling fraction and consists of a polymer of sulphated D-galactose units. Agar melts at about 100°C and gels at around 35°C. Agar is stable, does not react with media constituents and is not normally digested by plant enzymes.

Agarose

In plant tissue cultures, purified agarose (i.e. without agaropectin) is sometimes used instead of agar, particularly in circumstances where a high purity support matrix is required for example in protoplast and anther culture.

Gellan gums

Gellan gums such as Phytagel (Sigma) and Gelrite (Merck) are polysaccharides obtained from the bacterium *Pseudomonas elodea*. These gelling agents are polymers made up of glucose, glucuronic acid and rhamnose units. They produce clear, colourless gels that make visual screening for contamination easy. Gel strength is determined by the concentration of divalent cations (Mg^{2+} and Ca^{2+}). Poor gelling may occur if the cationic concentration is lower than 4 mM or higher than 8 mM.

Further reading

Lack, A.J. and Evans, D.E. (2001) *Instant Notes in Plant Biology.* Bios, Oxford.

Reference

Murashige, T. and Skoog, F. (1962) A revised medium for rapid growth and bioassays with tobacco tissue cultures. *Physiol. Plant.* 15: 473–497.

Chapter 3

Tissue culture in agriculture, horticulture and forestry

1. Introduction

Modern humans began farming about 10 000–12 000 years ago. This marked a shift from gathering food and fibre from wild plants, to the deliberate cultivation of crops. The transition was probably gradual but was one of the most dramatic changes in the history of mankind. As a result of settled agriculture, humans began to select, safeguard and propagate variants of species that had the most desirable phenotypic characteristics. The activities of planting, harvesting and storing seed for replanting and propagation imposed selection pressures. Over several millennia, these selection pressures resulted in changes in the phenotypes of wild plants, which led to the phenotypes that are characteristic of their domesticated crop relatives. With time, the farmer-selected crops became adapted to the local conditions and were maintained by man as 'primitive cultivars' or land races. Starting with land races, farmers over centuries used simplistic breeding methods to try to improve crops and to create plants that would grow effectively in regions of the world far beyond their centres of origin and domestication. However, the process of crop improvement by these methods was slow and it was not until the beginning of the 20th century that rapid progress was made, when sound scientific method was introduced to plant breeding by the application of Mendelian genetics. For the next four or five decades up until the early 1970s, methods of breeding for specific traits were developed that made significant contributions to agriculture, by delivering improvements in crop yield and enhancing plant resistance to pest and diseases.

Plant breeding depends on the sexual recombination of selected plants to generate new genotypes that can be propagated through seed. Over the centuries, in parallel with the evolution of plant breeding methods, techniques were developed for the asexual (vegetative) propagation of selected plants. In asexual propagation, the new plants are exact copies or clones of a single parent plant. Classical horticultural methods of clonal propagation include budding and grafting, cuttings, bulbs, corms, rhizomes and tubers.

From the early 1970s, as a result of improved methods of *in vitro* cultivation and new molecular biological procedures, plant cell and tissue culture has had a pronounced impact on both plant breeding and vegetative propagation (micropropagation) and has also simplified the storage and conservation of germplasm. In addition, plant tissue culture has been an integral part of the successful development of plant transformation technology, which has allowed the creation of transgenic plants by introducing DNA from almost any source.

The application of these procedures, known collectively as plant biotechnology, has become a versatile platform for basic scientific research as well as for commercial applications in agriculture, horticulture and forestry. This chapter will consider some of these applications and the following chapter (Chapter 4) will deal specifically with genetic modification.

2.   Micropropagation

Micropropagation is a tissue culture technique for producing large numbers of identical copies of a plant from a tissue fragment of a 'mother plant'. In the simplest case, meristematic tissue from mature plants is removed under aseptic conditions and placed in or on sterile medium containing minerals, essential vitamins and an appropriate combination of plant hormones (Chapter 2). The formation of the desired organs or tissues is induced by the action of the plant hormones. Usually, shoot development is induced on a medium designed for that purpose and the shoots are removed, separated and placed on a rooting medium (Chapters 10 and 11), where roots and complete plantlets will be formed. These plantlets are genetically identical, and the result is therefore clonal propagation.

Micropropagation has the immense advantage of rapidly generating a large number of genetically identical plants in a much shorter time than could be achieved by conventional propagation methods. It has particularly important applications for genetic modification, where a novel transgenic plant can be rapidly multiplied (Chapter 4).

Micropropagation has been applied successfully to the production of more than 1000 plant species. Many of these are ornamental, but the technique is also routinely applied to food crops and trees. It is particularly useful in several circumstances. The first is where it is troublesome to propagate a species vegetatively or where seed is recalcitrant (difficult to germinate). This includes ornamental species like the orchids. The second is where plant breeding is slow and introduces unwanted genetic variation, for instance in the case of tree species. The third is where the yield of a species is limited by a systemic disease that is transmitted by conventional propagation.

2.1   Commercial micropropagation

Commercial micropropagation of plants is used for the production of large numbers of copies of a selected genotype. As only very small pieces of the parent plant need be removed, the plant is not destroyed and so rare or unusual plants can be used as starting material. Tissue explants, which may be the shoot tip, lateral buds, leaf, root or stem, are placed on media that encourage large numbers of shoots to differentiate. These shoots are then separated and placed on rooting media to form plantlets. By repeated subculture of buds or shoots many plants can be produced *in vitro* all having the same genetic characteristics as the original plant. The plantlets must then be weaned from the axenic conditions in which they were formed into viable plants capable of survival in conventional horticultural or agricultural environments.

The operational steps in micropropagation are labour-intensive and repetitive and in some production systems gains in speed, sterility and labour cost have been achieved by the introduction of automation. The relatively high production costs of *in vitro* micropropagation compared with classical methods of propagation are compensated for by the following benefits:

- rapid propagation because of short propagation cycles;
- large volume propagation of high-value ornamentals like orchids, and of trees and shrubs;
- storage and transport of large numbers of plants;
- production of plants is independent of season;
- supply can react quickly to changes in demand.

Micropropagation techniques are widely used to generate clones of high-yielding or disease-resistant individuals. Some species are genetically variable and while this may have advantages in conferring a range of disease resistance, it limits productivity in commercial production. The problem is heightened where productivity can only be determined after a long period, for instance a tree or shrub that does not begin to yield a commercial crop for several years. Many trees (both gymnosperms and angiosperms) and plantation crops like bananas (*Musa* spp.) and oil palm (*Elaeis guineensis*) show high levels of genetic variation and have been particularly suitable for the clonal propagation of plants selected for their agronomic traits (superior plants).

One of the first cultivated species to be improved in this way was the oil palm. Oil palms, which are grown extensively in Africa, Malaysia and Indonesia, account for 20% of world oilseed production. Yield can vary by as much as 30% per tree and 30 million trees are replaced annually. Conventional methods of vegetative propagation (cuttings, shoots or grafts) cannot be used for this plant with its obligate cross-pollination and whose natural means of propagation is by seed. Because of the substantial heterogeneity of plants grown from seed, desirable agronomic traits can only be fixed by clonal selection and propagation. Large-scale clonal propagation of selected plants is achieved by regeneration using somatic embryogenesis. The oil palm propagation programme has concentrated on yield improvement, ease of mechanical harvesting and disease resistance.

Similar principles of selection are used for the micropropagation of many other crops. Members of the plantain and banana families, for instance, are major crops of the humid tropics with an annual production of more than 85 million tonnes (FAO, 1999). Of this, around 10% is for export, the remainder being a staple for indigenous populations. *Musa* crops present a range of problems for improvement by conventional breeding methods. They are generally triploid and therefore sterile; they take 2 years from seed to seed and field trials require large amounts of space. Micropropagation of *Musa* based on shoot tips allows production of large numbers of pest- and disease-free plants. Together with other methods, it has permitted the production of faster-growing and higher-yielding plants that are popular with producers (Crouch *et al.*, 1998).

Other species commercially produced by micropropagation include: orchids, other flowers and ornamental plants, date (*Phoenix dactilyfera*), soft coconuts (*Cocos nucifera*), cardamom (*Elettaria cardamomum*), eucalyptus (*Eucalyptus globulus*), Chinese fir (*Cunninghamia lancolata*), rattan (*Calamus* spp.) and triploid water melons (*Citrullus lanatus*).

2.2   Production of virus-free plants

Plant viruses are systemic and transmitted by conventional propagation methods. The discovery that the meristem is often free of infection resulted in the first production of virus-free *Dahlia* in the 1950s. By the end of the decade, the technique had been used to remove paracrinkle virus from potato (*Solanum tuberosum*). The technique has subsequently been successfully applied to a wide range of crops and ornamental species. These include: banana, rhubarb (*Rheum raponticum*), strawberry (*Fragaria* spp.), citrus species, papaya (*Carica papaya*), apple (*Malus* spp.), sweet cherry (*Prunus aidum*), pear (*Pyrus communis*), grapes (*Vitis vinifera*), sweet potato (*Ipomoea batatas*), *Allium* spp., taro (*Colocasia esculentus*), sugar cane (*Saccharum officinarum*) and ornamentals such as chrysanthemum (*Chrysanthemum coronarum*), carnation (*Dianthus caryophyllus*) and *Lilium* spp. It is likely that meristems remain virus free because they lack direct vascular connections to the rest of the plant and are continuously growing away from older, infected tissues. It may also be the case that virus movement is limited through plasmodesmata adjacent to the meristem.

To produce virus-free plants, virus-free meristematic tissue in shoot-tips or axillary buds of infected plants are cultured and regenerated to produce new plants. For success, the size of the explant, which varies with species, is the crucial factor. It must be as small as possible in order to eliminate all virus-infected tissue, but large enough for regeneration. If necessary, tissues may be heat-treated or incubated with an anti-viral agent like malachite green or thiouracil to kill remaining virus. A modification of this technique, known as micrografting, is used to produce virus-free woody fruit trees. In this procedure, sterile virus-free axillary buds are grafted onto a rootstock grown under sterile conditions. Potato and strawberry are examples of commercially grown crops that depend on a supply of virus-free stock each season.

2.3   Somatic embryogenesis and propagules

In the tissue culture techniques described so far, plantlets are frequently produced by meristem culture. As an alternative, embryos may be generated by cell and tissue culture (somatic embryos), either by the use of an embryogenic callus (Chapter 6) or from suspension cultures (Chapter 7). Suspension cultures may be grown on a large scale and thousands of embryos generated. The embryos formed are genetically identical and can be used as if they were zygotic (seed) embryos. However, the seed is a complex nutritional and protective system that greatly enhances embryo survival and facilitates propagation. Handling naked somatic embryos presents technical difficulties especially in commercial agriculture and horticulture where mechanized sowing of seeds is conventionally used. To overcome this problem, artificial propagules (or pseudo-seeds) have been developed, by surrounding the embryo with a hydrated gel. They have been produced for a variety of ornamental species, including lilies, azaleas and African violets, as well as for some trees and crops like tomato, asparagus (*Asparagus officinalis*)

Table 3.1. Forest trees micropropagated by the Canadian Forest Service (nrcan, 2002)

White spruce	*Picea glauca* (Moench) Voss
Black spruce	*Picea mariana* (Mill.) BSP
Red spruce	*Picea rubens* Sarg
Tamarack	*Larix laricina* (Du Roi) K. Koch
European larch	*Larix decidua* Mill
Hybrid larch	*L. decidua* Mill. × *Larix leptolepis* (Siebold & Zucc.) Gord.
Eastern white pine	*Pinus strobus* L.
Jack pine	*Pinus banksiana* Lamb.

and celery (*Apium graveolens*). They can be successfully used for mechanical sowing.

Micropropagation based on somatic embryogenesis is also used worldwide for the commercial propagation of many major crops, including rubber, sugar cane, potato and plantain. It is also widely applied in forestry and agroforestry. Trees like eucalyptus, poplar, bamboo, teak, spruce and pine are all commercially micropropagated in areas in which they are grown commercially (FAO, 1995). *Table 3.1* lists some of the forest trees currently produced by this means by the Canadian Forest Service.

3. Plant breeding

Over the last 25 years, tissue culture methods in combination with other procedures have been applied to plant breeding. In many breeding programmes, it is now possible to use classical methods, new techniques or a combination of classical and new. In this section, we describe the tissue culture based methods and illustrate how their applications can be combined with other techniques. Sources of genetic variation for plant improvement are summarized in *Figure 3.1*.

3.1 *In vitro* selection

The concept of *in vitro* selection is to exploit the genetic variation known to occur in plants by screening cell cultures for resistance to disease, insects, herbicides or stress. The procedure of *in vitro* selection typically involves subjecting cells in culture to a suitable selection pressure and recovering any variant cell lines that are resistant to the stress. These variant lines are then used to regenerate whole plants. The technique relies on a dependable method of regeneration from callus and presupposes that resistance displayed by undifferentiated cells in culture is equally expressed in whole plants and can be transferred to progeny by conventional plant breeding.

The use of *in vitro* screening to select for herbicide resistance is particularly effective as the herbicide can be added to the culture in defined concentrations. However, the mode of action of the herbicide is also an important factor, as it must be the same at cellular level as for the whole plant. Herbicide resistance has been selected by *in vitro* screening, for a range of herbicides in a number of plants. For example, glyphosate resistance in carrots was selected by subjecting carrot suspension cultures to repeated subcultures with incremental increases of glyphosate concentration in the

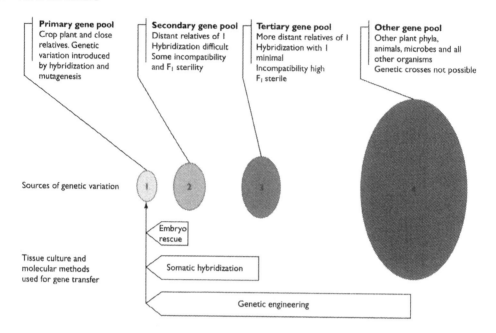

Figure 3.1

A general scheme summarizing the sources of genetic variation available for the improvement of crop plants and the methods that can be used for gene transfer to the crop. In this scheme, the segregation of gene pools is based on relative ease of hybridization.

media (Shyr and Widholm, 1990). Regenerated plants in the field were resistant to glyphosate.

In vitro screening for resistance to plant pathogens is most successful when the disease is caused by known phytotoxic compounds produced by the pathogen. Screening for resistance to the compounds or their analogues has led to the selection of plants resistant to a number of bacterial and fungal diseases, for example resistance to *Helminthosporium oryzae* in rice (Vidhyasekaran *et al.*, 1990).

The frequency of genetic variation in plant populations can be increased by mutagenesis and mutagenized material can be subjected to *in vitro* screening to select desired variants. Mutagenesis can be brought about by:

- ionizing radiation;
- chemical compounds;
- cell and tissue culture regimes (somaclonal variation).

Mutagenesis induced by chemicals and ionizing radiation

There are several chemicals used to induce mutagenesis in the genomes of plants. The most used chemicals are alkylating agents, and the most preferred alkylating agent is the compound ethyl methane sulphonate (EMS). The ionizing radiation may be neutrons, beta-radiation or X-rays. Seeds are mutagenized, germinated and progeny selected for the desired trait. Progeny may be rapidly micropropagated and backcrossed with other lines to combine the useful trait of the mutant with desirable existing features of the crop. About 2000 crop cultivars have been generated by mutagenesis and the FAO lists such cultivars available for crop breeding (FAO, 2002). Many current lines of crops, like rapeseed (*Brassica napus*), wheat (*Triticum* spp.), white bean (*Phaseolus vulgaris*) and barley (*Hordeum vulgare*), are the result of crosses with mutagenized lines. Thus, while mutagenesis has diminished as interest in plant genetic engineering has increased, it still remains an important technology.

Somaclonal variation

When an apparently genetically identical clone of plants, regenerated from cell and tissue cultures, is grown to maturity, some individuals show clear variation in characteristics – size, yield, colour etc. Some of these variants are useful; others are deleterious. These aberrants are known as somaclonal variants and may be used as part of a selection-breeding programme in crop improvement. Somaclonal variation leads to the creation of additional genetic variability. Somaclonal variation was used extensively in the 1980s and 1990s in plant breeding. The rate of variation exceeds that of natural mutation and results from the impact of the tissue and cell culture technique on the genome of plants being micropropagated. Most of the variations can be attributed to chromosomal instability and often the degree of instability is correlated with the length of time the cells were in culture. Since the early studies of Larkin and Scowcroft (1981) suggested that somaclones were a novel source of variation that could be exploited for crop improvement, the procedure has been applied to a wide variety of species. There are several examples where somaclonal variation was used to improve crop plants. These included melon (*Cucumis melo*), strawberry and rice (*Oryza sativa*) (disease resistance, food quality and salinity resistance, respectively), sugar cane (disease resistance), banana, wheat, barley, maize, tomato (disease resistance) and brassicas (development and maturation). Very few somaclonal variants are now being used commercially due to the genetic instability of the variants, and this strategy for crop improvement is now questioned (see Chapter 12). Somaclonal variation is also a major disadvantage when clonal uniformity is required, for example where tissue culture is employed for propagation of elite genotypes.

3.2 Somatic hybridization

A novel and elegant use of cell and tissue culture technology for plant breeding is the production of somatic hybrids by protoplast fusion. Protoplasts are single plant cells that have had their cell walls removed by enzymatic digestion (see Chapter 8). Protoplasts can grow new cell walls to produce intact cells, each theoretically capable of regeneration to a whole plant.

The concept of somatic hybridization was based on the observation that when protoplasts are brought into close contact, sometimes they would fuse with each other. Subsequently, it was found that fusion could be greatly enhanced either by applying a brief electrical pulse or by treatment with compounds like polyethylene glycol (PEG).

The fusion of protoplasts from different species produce so called somatic hybrids. Somatic hybridization is an *in vitro* technique that makes it possible to circumvent the incompatibility systems that prevent *in vivo* hybridization between sexually incompatible species. The first successful fusion was between two species of *Nicotiana* (Carlson *et al.*, 1972), and reports of other fusions soon followed, including the somatic hybrid from tomato and potato (Melchers *et al.*, 1978).

Early expectations that somatic hybridization would play an important role in the development of new varieties have remained unfulfilled. However, some notable applications have resulted from somatic hybridization in closely related species. Fusion between *Nicotiana tabacum* and *Nicotiana rustica* has produced tobacco plants that varied in nicotine and tar contents, as well as their resistance to blue mould and black root rot (Pandeya *et al.*, 1986). Somatic hybridization has been used extensively with citrus, for instance to produce interspecific hybrids between the two sexually incompatible species *Citrus reticulata* and *Citropsis gilletiana* (Grosser *et al.*, 1990). Hybrids have also been produced between apple species (Saito *et al.*, 1989).

The fusion of haploid protoplasts has been used to shorten breeding programmes, and fusions between the nuclear genome of one species and the cytoplasmic genome (mitochondria and chloroplasts) of another have produced a fusion product called a cybrid (**cytoplasmic hybrid**). Cybrids are of particular interest in producing cytoplasmic male sterile (CMS) lines for hybrid breeding, as the CMS trait is carried by the mitochondria. CMS cybrids of rice have been produced and used in hybrid seed production (Kyozuka *et al.*, 1989) and similar cybrids have been produced for other crops (FAO, 1995).

3.3 Haploidy

The production of haploid plants (plants having only one set [n] of chromosomes) is one of the major contributions of plant tissue culture to plant breeding (see Chapter 9). Haploid plants can be induced to double their chromosome number [$2n$] by chemical treatment, producing double haploids (dihaploids) that are homozygous for all genes. Dihaploids are desirable as parents for breeding of F_1 hybrids and facilitate the selection of recessive traits. Tissue culture techniques have simplified breeding programmes by dramatically reducing the time required to produce haploids. There are two commonly used *in vitro* methods for the production of haploids:

- anther culture;
- the culture of microspores.

In both cases, embryogenesis is induced in the cultures and the embryos used to regenerate whole plants (Chapter 9).

Haploids have been used in plant breeding to increase yield and improve traits like time to maturity and resistance to abiotic stress. There are examples of cultivars resulting from haploid breeding in many species; China has been particularly active in this area, with new varieties of rice, wheat, tobacco and hot and sweet peppers (*Capsicum* spp.). It was estimated that in 1995, 1.5 million ha, mostly of rice and wheat produced by haploid techniques, were under cultivation (FAO, 1995). Elsewhere, rice, Chinese cabbage

(*Brassica rapa*), broccoli (*Brassica oleracea*), barley, maize, tobacco, strawberry and asparagus have been generated by haploid breeding.

3.4   Embryo rescue and culture

Embryo culture was one of the first successful plant tissue culture techniques attempted. In 1904, Hannig cultured mature embryos of the crucifers *Raphanus* and *Cochleria* using aseptic techniques. The principal application of embryo culture is to rescue embryos produced as the progeny from interspecific (wide) crosses that normally fail to develop viable seed. In these crosses, post-zygotic incompatibility mechanisms caused early abortion of embryos. This barrier to embryo development can be overcome by removing the immature embryo from the developing seed and culturing it *in vitro* to produce the hybrid plant. Wide crosses and embryo rescue are important tools for the transfer of desired traits such as disease resistance, stress tolerance or high yield from wild species to crop species.

4.   Biodiversity and conservation of germplasm

The intensification of agricultural production seen during the second half of the 20th century with the emphasis on increasing output by the large-scale cultivation of high-yielding elite clones, has reduced both the number and the genetic diversity of the species used in agriculture. Crop uniformity and neglect of the genetic diversity found in wild related species has had the effect of narrowing the gene pool from which parental material has been drawn. The result of the narrowing of the genetic base is an increased vulnerability of crops to pests, diseases and adverse abiotic factors. There is an increasing awareness that the genetic resources of wild and cultivated plants together constitute a wide pool of agronomically valuable genes that will be needed in the decades ahead, to increase tolerance of drought and poor-quality soils, boost resistance to diseases, maximize yields and improve nutritional quality. In response to this problem, there has been a worldwide effort to collect and conserve plant genetic resources and to develop techniques for the long-term storage and maintenance of germplasm.

The are three main categories of genetic resources of crop plants that are conserved:

- germplasm from the wild relatives of crop plants;
- germplasm derived from traditional breeding programmes;
- biotechnologically derived plant germplasm.

Although field gene banks have been an important means of conservation, seed storage has been the most widely used method of conserving germplasm. Conventionally dehydrated seeds are stored at low temperatures, in cold rooms. However, many species cannot be conserved by this method because their seeds do not tolerate dehydration and/or cold (e.g. coffee, *Coffea* spp.), or they are propagated vegetatively (e.g. banana). Plant tissue culture has provided an alternative system for both *in vitro* collecting and *in vitro* conservation of germplasm.

4.1   *In vitro* collecting

Essentially, the technique of *in vitro* collecting involves taking an explant in the field, decontaminating the material and placing it in sterile culture medium before transfer to a plant tissue culture laboratory for further *in vitro* processing (e.g. Hamill *et al.*, 1993). This technique is particularly effective for species with recalcitrant seeds or embryos that deteriorate rapidly.

4.2   *In vitro* culture and germplasm storage

In vitro culture offers the advantage that disease-free cultures can be established from explants and stored under optimal conditions for long periods. For instance, virus infection can be eliminated by meristem culture as described earlier (Section 2.2). In addition, germplasm stored as disease-free cultures can be easily transferred between countries without the need for quarantine restrictions.

In vitro culture collections may consist of dedifferentiated cell cultures such as suspension and callus culture, or organized tissue cultures like embryos or meristems. Cultures are maintained either in slow-growth storage or by cryopreservation.

Slow-growth storage reduces the frequency of subcultures and saves on labour and media costs. Slow growth is achieved by a variety of methods including low temperature (2–8°C), low light, low oxygen, desiccation or culture on minimal medium with growth retardants.

Cryopreservation is the storage of cultures at or near the temperature of liquid nitrogen (−196°C). At this temperature, all cellular processes are effectively stopped and germplasm can be stored for very long periods without risk of contamination. Cryopreservation is achieved by a number of different procedures that either use cryoprotectants and controlled cooling or apply dehydration followed by rapid freezing (Benson *et al.*, 1998).

The techniques of *in vitro* storage go beyond that of preserving germplasm for plant breeding; they also offer possibilities of maintaining material of rare and endangered species.

References

Benson, E.E., Lynch, P.T. and Stacey, G.N. (1998) Advances in plant cryopreservation technology: current applications in crop plant biotechnology. *AgBiotech News and Information*, Vol. 10, No. 5, 133N–142N. AgBiotechnet.com

Carlson, P.S., Smith, H.H. and Dearing, R.D. (1972) Parasexual interspecific plant hybridization. *Proc. Natl Acad. Sci. USA* 69: 2292–2294.

Crouch, J.H., Vuylsteke, D. and Ortiz, R. (1998) Perspectives on the application of biotechnology to assist the genetic enhancement of plantain and banana (*Musa* spp.) *Electronic J. Biotechnol.* 1: 11–22.

FAO (1995) Research and Technology Paper 6. ISBN 92-5-103626-8, http://www.fao.org/docrep/V4845E/V4845E00.htm

FAO (1999) Committee on commodity problems. Intergovernmental group on bananas and tropical fruit. Projections for supply and demand of bananas to 2005. ftp://ext.ftp.fao.org/waicent/pub/ccp/bntf99/x1065e.pdf

FAO (2002) http://www-mvd.iaea.org/MVD/default.htm

Grosser, J.W., Gmitter, F.G., Tusa, N. and Chandler, J.L. (1990) Somatic hybrid plants from sexually incompatible woody species: *Citrus reticulata* and *Citropsis gilletiana. Plant Cell Rep.* 8: 656–659.

Hamill, S.D., Sharrock, S.L. and Smith, M.K. (1993) Comparison of decontamination methods used in initiation of banana tissue cultures from field-collected suckers. *Plant Cell, Tissue Organ Cult.* 33: 343–346.

Kyozuka, J., Kaneda, T. and Shimamoto, K. (1989) Production of cytoplasmic male sterile rice (*Oryza sativa* L) by cell-fusion. *Bio-Technology* **7**: 1171–1174.

Larkin, P.J. and Scowcroft, W.R. (1981) Somaclonal variation – a novel source of variability from cell cultures for plant improvement. *Theor. App. Genet.* **60**: 197–214.

Melchers, G., Sacristan, M.D. and Holder, A.A. (1978) Somatic hybrid plants of potato and tomato regenerated from fused protoplasts. *Carlsberg Res. Commun.* **43**: 203–218.

nrcan (2002) http://www.nrcan.gc.ca/cfs/proj/sci-tech/biotechnology/treepr_e.html

Pandeya, R.S., Douglas, G.C., Keller, W.A. *et al.* (1986) Somatic hybridization between *Nicotiana rustica* and *Nicotiana tabacum* – development of tobacco breeding strains with disease resistance and elevated nicotine content. *Z. Pflanzenzuchtung – J. Plant Breeding* **96**: 346–352.

Saito, A., Niizeki, M. and Saito, K. (1989) Organ formation from calli and protoplast isolation, culture, and fusion in apple, *Malus pumila* Mill. *J. Jpn. Soc. Horticult. Sci.* **58**: 483–490.

Shyr, Y.Y. and Widholm, J.M. (1990) Glyphosate resistance and gene amplification in selected *Daucus carota* suspension cultures. In: *Progress in Plant Cellular and Molecular Biology* (eds H.J.J. Nijkamp, L.H.W. Van Der Plas and J. Van Aartrijk). Kluwer, Dordrecht, 148–152.

Vidhyasekaran, P., Ling, D.H., Borromeo, E.S. *et al.* (1990) Selection of brown spot-resistant rice plants from helminthosporium-oryzae toxin-resistant calluses. *Ann. Appl. Biol.* **117**: 515–523.

Chapter 4

Tissue culture in genetic engineering and biotechnology

1. Introduction

In addition to the many applications described in Chapter 3, plant tissue culture is also a prerequisite for the genetic engineering of transgenic plants with novel characteristics. The technical operation of introducing and expressing foreign genes in plants was first described for tobacco in 1984 (De Block *et al.*, 1984). Since then this technology has been extended to over 120 species and large areas of transgenic crops are now cultivated worldwide.

For the greater part of the 20th century, plant breeding was the traditional method of generating cultivated plants with desired characteristics. Selective plant breeding (hybridization) brings together desired genes from two or more individuals with a resultant combination of desired traits in the offspring – a hybrid. Selective plant breeding has combined applied Mendelian genetics with other methods, such as the induction of mutations, to achieve a number of specific breeding aims. For example, crop yields have been increased, qualitative traits, such as palatability and aesthetic features, have been improved and resistance to disease and pests introduced.

The creation of new genotypes by classical plant breeding largely depends on mechanisms that occur during meiosis in germ cells. In contrast, modern biotechnological methods bypass the generative cycle and produce new genotypes by introducing desired genes directly into somatic cells (*Figure 4.1*). Furthermore, the development of tissue culture and molecular biological procedures for the exchange of DNA between unrelated organisms has enabled novel genes derived from organisms outside the plant kingdom to be introduced into recipient plants. In addition, this modification of plant genomes by the integration and expression of foreign DNA (genetic engineering) leads to the formation of new genotypes (transgenic plants) that can be developed further by classical plant breeding methods.

2. Genetic engineering

Four key steps are required to produce a transgenic plant.

(i) Identification and isolation of the gene(s) of interest from the original organism.

(a)   (b)

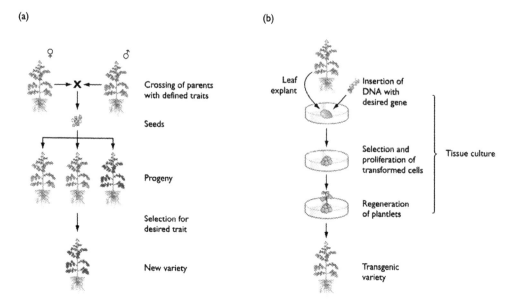

Figure 4.1

A comparison of the steps required to produce a new variety by traditional plant breeding (a) and a transgenic variety by genetic engineering (b).

(ii) Integration of the isolated gene into a plasmid and multiplication in bacteria (cloning).
(iii) Insertion of the cloned gene into cells of a target plant.
(iv) Selection and growth of transformed cells into new plants.

Each step requires specific tools and procedures.

2.1 Identification and isolation of the desired gene

The essential first step in genetic engineering is finding and making copies of the specific gene that encodes the protein of interest. A range of techniques is available for locating and sequencing stretches of DNA that contain a desired gene. Commonly, these procedures include identification of the desired gene through screening of a DNA (gene) library of the donor organism. Screening is based on detecting the DNA sequence of the cloned gene, detecting the protein that the gene encodes or by using linked DNA markers. The identified gene is joined by ligation into a bacterial plasmid (a cloning vector) that is copied (cloned) by DNA replication in the host bacterium. Frequently, the gene of interest is cloned from a genomic DNA library and subcloned from a cDNA library. The cDNA form of the gene is preferred to the genomic DNA form, as cDNA transcripts will require no RNA processing before translation. This is particularly important in heterologous expression, as the host plant may not have the appropriate machinery to produce functional messenger transcripts.

2.2 Construction of transgene and integration in cloning vector

The isolated gene is integrated into a bacterial plasmid (a cloning vector) and multiplied in a host bacterium like *Escherichia coli*. The cloned gene must be modified so that it will be effectively expressed when inserted into the plant. Typically the modifications involve changes to sequences in the region of the gene that control gene expression, however sometimes the cloned gene itself is modified to achieve greater expression in a plant. For example, changes can be made to overcome codon bias. Plants prefer G–C nucleotide pairs compared to bacterial genes, which have a high percentage of A–T nucleotides. In cloned bacterial genes, A–T nucleotides can be substituted for G–C nucleotides without significantly changing the amino acid sequence, but enhancing the production of the gene product.

Promoters and gene expression

In addition to suitable start and termination sequences, each construct must contain an appropriate promoter. The promoter ensures that the protein encoded by the gene is produced either constitutively, during a particular phase of development or in specific organs. The most commonly used promoter in plant transformation is the 35S promoter (35S CaMV). The 35S CaMV promoter is a constitutive promoter derived from the DNA-based cauliflower mosaic virus (CaMV) and is used to regulate the expression of transgenes in several transgenic crops currently under cultivation. The 35S CaMV promoter can function with both monocots and dicots and maintains a high and constant level of transcription.

In many genetic engineering experiments, tissue- or developmental-specific expression is an important requirement. Consequently, a great effort is currently being made to identify plant promoters of known specificity. A tissue-specific plant promoter that is currently used in transgenic crops is the phosphoenolpyruvate (PEP) carboxylase promoter. The PEP carboxylase promoter is from the plant gene encoding a photosynthetic enzyme and this promoter will be active only in cells that are producing photosynthetic proteins. Expression also slows and stops at the end of the growing season when the requirement for photosynthesis has diminished.

Selectable markers and reporter genes

In addition to the desired gene, transformation vectors are constructed with a second gene, either a selectable marker gene or a reporter gene. These genes have no designated function either in the transformed plant or in any products derived from it, but are necessary to identify cells or tissues that have successfully integrated the transgene. This is essential because the frequency of achieving incorporation and expression of transgenes in plant cells is generally limited to a few per cent of the targeted tissues or cells. Selectable marker genes and reporter genes are inserted in the same stretch of DNA as the gene of interest where they are fused to the same promoter and they require termination sequences for proper function (*Figure 4.2*).

Selectable marker genes encode proteins that provide resistance to agents that are normally toxic to plants. Only those cells that have integrated and expressed the selectable marker gene will survive when grown on a medium containing the appropriate toxic compound. In contrast, reporter genes do not affect survival on toxic media but code for gene products that have easily detectable phenotypes.

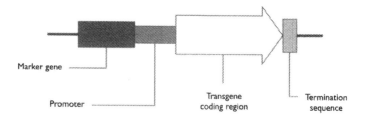

Figure 4.2

A simplified scheme showing the regions (and their relative positions) of a transgene construct that are required for integration and expression.

Selectable marker genes are almost exclusively of two types: (i) genes conferring antibiotic resistance and (ii) genes conferring herbicide resistance.

There are several commonly used antibiotic resistance markers. Currently, the most widely used antibiotic resistance marker is the neomycin phosphotransferase type II gene (*nptII*), which is derived from the bacterial transposon Tn5. Neomycin phosphotransferase II (NPTII) is an enzyme that confers resistance by inactivating a number of related antibiotic aminogly-cosides, such as kanamycin, paromomycin and geneticin. Transformed cells expressing NPTII are protected from the effects of the kanamycin and can grow and regenerate into whole transgenic plants in cell culture media containing the antibiotic. A number of other antibiotic resistance markers are employed in the production of transgenic plants. These include the *aad* gene for the enzyme 3'(9)-O-aminoglycoside-adenylyltransferase, that confers resistance to streptomycin and spectinomycin and the *hpt or hph* gene from *E. coli*, which confers resistance to hygromycin. The resistance gene codes for a kinase that inactivates hygromycin through phosphorylation.

Two herbicide selectable markers are commonly used in plant transformation.

The first is the *bar* gene encoding the enzyme phosphinothricin acetyl-transferase (PAT), isolated from *Streptomyces*. PAT confers resistance to the herbicide phosphinothricin by catalysing its detoxification to the acetylated form. The second herbicide marker gene is the mutant allele of the aroA locus of *Salmonella typhimurium* encoding a 5-enolpyruvylshikimate-3-phosphate (EPSP) synthase in which a single substitution of a proline for a serine causes a decreased affinity for the herbicide glyphosate (Roundup), without affecting the kinetic efficiency of the enzyme. Non-transformed cells can survive but have an altered phenotype.

Reporter genes

After transformation, there are several ways of determining where and when the gene of interest is expressed in the host plant tissues. Some of these procedures, such as Southern or northern hybridization, dot blot analysis, enzymatic assays or polymerase chain reaction (PCR) are either laborious or require careful optimization. Reporter genes are a rapid and convenient strategy to characterize the expression patterns of the transgene. With most reporter genes, successful transgene expression can be determined through a visual assay.

Often, the reporter gene encodes a protein that can be easily detected either by characteristics such as fluorescence (fluorogenic) or through an enzyme activity that produces a coloured product (chromogenic). Thus, the location and amount of gene expression in a transformed tissue can be readily assessed.

The β-galactosidase (lacZ) and β-glucuronidase (GUS) genes are two examples of reporter genes that are frequently used. An important requirement of reporter genes is that their activity is absent in the plant in which they will be used. Both lacZ and GUS are genes derived from *E. coli*; however, some plants do contain some β-galactosidase activity and this can lead to high background staining. GUS activity is normally very low in plants, and so it is a common reporter gene used in plant studies. GUS activity is visualized by transferring the transformed tissue to a medium containing 5-bromo-4-chloro-3-indoyl-1-glucuronide, which is a substrate for the β-glucuronidase enzyme. Cleavage of the substrate by β-glucuronidase produces a blue-coloured product. The detection of some of the most common reporter genes used in plant transformation is shown in *Table 4.1*. The expression pattern determined by the reporter gene can then be confirmed by one of the other more quantitative strategies like Southern analysis.

2.3 Genetic transformation

Transformation is the insertion and integration of the DNA representing the cloned gene into the genome of the host plant cell so that it expresses the protein encoded by the gene. It is important that the foreign DNA is integrated into the host plant chromosomes so that it can be passed on during mitosis to future cell generations. This is known as 'stable transformation' and transformed cells can be used to regenerate a complete plant where every cell contains a copy of the foreign gene. If the DNA of the inserted gene is not integrated into a chromosome, it can still be expressed for a short time, a phenomenon known as 'transient expression'.

The insertion of genes into plant cells can be achieved by the following two types of procedures:

(i)   Indirect gene transfer that exploits the naturally occurring vector *Agrobacterium tumefaciens*.
(ii) Direct gene transfer, which is based on the physical methods that drive the uptake of DNA.

Indirect gene transfer

A. tumefaciens is a Gram-negative soil bacterium that infects susceptible plants causing the formation of tumorous growths that are characteristic of a pathological condition known as crown gall disease. The bacterium attaches to wounded tissue at the base of the stem and induces multiplication of the underlying cortical cells. During the process of infection, an independent loop of *Agrobacterium* DNA (more than 200 kb) known as the 'Tumour-inducing' or Ti-plasmid is inserted into the host cell. A mobile DNA segment of the Ti-plasmid (known as T-DNA) is transferred to the nucleus of infected cells where it integrates into the host's genome and is transcribed like a normal part of the plant's DNA.

T-DNA contains two types of genes: oncogenic genes encoding the enzymes responsible for the biosynthesis of phytohormones (auxins and

Table 4.1. Activities and measurement of proteins encoded by reporter genes used in plant transformation

Protein	Activity	Measurement
GUS (β-D-glucuronidase)	Hydrolysis of β-D-glucuronides	(i) With the chromogenic substrate 5-bromo-4-chloro-3-indolyl-β-D-glucuronide, the product formed (indoxyl derivative) is oxidized to a blue precipitate. (ii) With the fluorogenic substrates 4-methylumbelliferyl-β-D-glucuronide (MUG). Fluorescence can be measured at an excitation of 360 nm and an emission of 450 nm.
LacZ (β-D-galactosidase)	Hydrolysis of β-D-galactosides	(i) With the chromogenic substrate 2-nitrophenyl β-D-galactopyranoside. The product is 2-nitrophenol (yellow in alkaline solution) (ii) With fluorogenic substrate 3-carboxyumbelliferyl β-D-galactopyranoside (CUG). The fluorescent product is measured at an excitation of 390 nm and an emission of 460 nm.
LUC (luciferase)	Oxidation of luciferin	Luminescence detection (photon emission)
GFP (green fluorescent protein)	Fluorescence	Irradiation with blue light

cytokinins) that are responsible for tumour formation; and genes encoding proteins involved in the synthesis of a group of compounds known as opines that are utilized by the bacteria as sources of nitrogen and carbon. The T-DNA is flanked by 25-bp imperfect repeats that define the boundaries of the T-DNA and act as a *cis* element signal in the transfer and integration process. T-DNA transfer is mediated by the combined action of genes encoded by the virulence region (35 *vir* genes) of the Ti-plasmid. Wounded cells of host plants secrete low molecular weight phenolic compounds (acetosyringone and hydroxyacetosyringone) that stimulate the *vir* genes

Agrobacterium-mediated transformation exploits this natural transformation mechanism. The genes within the T-DNA region are not required for DNA transfer and can be removed and replaced by desirable genes making it possible to use the Ti-plasmid as a transformation vector without disease induction. Transformation vectors and *Agrobacterium* host strains that are no longer oncogenic (i.e. they are disarmed) have been developed for plant transformation.

Transformation vectors

A primary requirement for *Agrobacterium*-mediated transformation is inserting the gene of interest into the T-DNA in preparation for transfer to the plant cell. Two classes of transformation vectors have been developed to achieve this aim.

- Integrating (or cointegrate) vectors, where the gene to be transferred is engineered into T-DNA on a disarmed Ti-plasmid containing *vir* genes.
- Binary vectors, where T-DNA modified to carry the desired gene and the *vir* region reside on separate plasmids.

Binary vectors are now the most commonly used strategy as they are easier to manipulate.

Agrobacterium transformation is a commonly used method of transformation for dicots because of the flexibility and ease of the procedure. In addition, single-copy integrations are the most common event, thus reducing potential problems of co-suppression and instability of the transgene.

Plant transformation is achieved by incubating (co-cultivating) *Agrobacterium* with a suitable explant. The bacterium is killed with an antibiotic and transformed cells are selected on an appropriate screening medium. Selected cells are allowed to proliferate in tissue culture and used to regenerate transgenic plants. Explants that are typically used for transformation in dicots include leaf discs, stem segments, cotyledons, callus suspension cultures, protoplasts and germinating seeds.

In some species, a number of factors are found to be crucial for successful transformation. These include the *Agrobacterium* strain and the temperature and duration of co-cultivation. *Agrobacteria* are attracted to wounded plant cells and pre-wounding of tissue with glass beads, microprojectiles, sonication or the addition of hydroxyacetosyringone has been used to increase transformation efficiency in some species.

It was previously thought that monocots were recalcitrant to *Agrobacterium*-mediated gene transfer because of a barrier to T-DNA integration. However, recently reliable and efficient procedures have been established for rice, maize, wheat, barley, banana and sugar cane.

Direct gene transfer

A major obstacle for all gene transfer systems including biological delivery systems is the impenetrability of the plant cell wall. Direct gene transfer techniques attempt to overcome this problem by using physical methods to facilitate the uptake of naked DNA by plant tissues.

The gene gun The most frequently used method of direct gene delivery is termed biolistic or ballistic transformation. This technique uses the mechanical force generated by high-velocity particles (microprojectiles) to propel DNA through the biological barriers of plant cells. In essence, particles of gold or tungsten (0.2–0.3 µm) are coated with DNA and shot into target cells by acceleration. The acceleration is generated by a particle gun using gunpowder, high-pressure gases (helium, carbon dioxide or nitrogen) or an electric discharge (*Figure 4.3*). This technique has the potential to deliver DNA into any tissue from any species of monocots and dicots, and to all

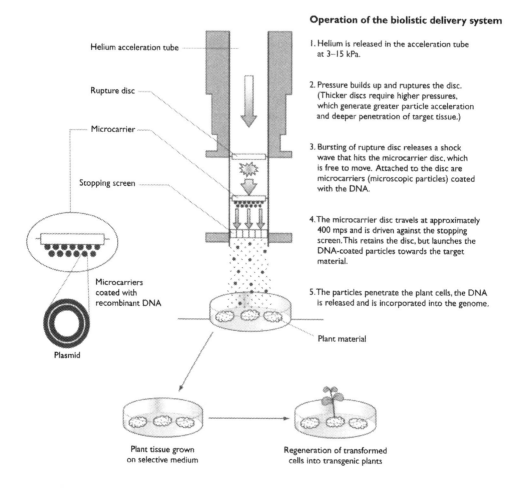

Operation of the biolistic delivery system

1. Helium is released in the acceleration tube at 3–15 kPa.

2. Pressure builds up and ruptures the disc. (Thicker discs require higher pressures, which generate greater particle acceleration and deeper penetration of target tissue.)

3. Bursting of rupture disc releases a shock wave that hits the microcarrier disc, which is free to move. Attached to the disc are microcarriers (microscopic particles) coated with the DNA.

4. The microcarrier disc travels at approximately 400 mps and is driven against the stopping screen. This retains the disc, but launches the DNA-coated particles towards the target material.

5. The particles penetrate the plant cells, the DNA is released and is incorporated into the genome.

Helium acceleration tube

Rupture disc

Microcarrier

Stopping screen

Microcarriers coated with recombinant DNA

Plasmid

Plant material

Plant tissue grown on selective medium

Regeneration of transformed cells into transgenic plants

Figure 4.3

Structure and operation of the gene gun used for microprojectile bombardment.

genomes in the plant cell including those of the mitochondrion and chloroplast. In some instances, pre-treatment of the tissue before bombardment has improved transformation. For example, proper pre-culturing of the explant tissue by an osmotic adjustment delivered by partial drying or by the addition of an osmotic agent to the medium.

Alternative methods Several alternative methods of direct gene transfer have been devised but for practical reasons they are not often used. Polyethylene glycol (PEG), an agent known to induce fusion of biological membranes has been used to promote the uptake of isolated DNA into plant protoplasts. Other ways of facilitating the uptake of naked DNA by plant cells include electroporation, laser microbeams, silicone carbide whiskers and microinjection with small cannulas.

2.4 Recovery and regeneration

Irrespective of the transformation system used, successful development of a transgenic plant depends critically on the ability of the transformed tissue to produce totipotent cells that proliferate and regenerate into a complete viable plant (*Figure 4.4*). The selection of successful transformants, and their regeneration and propagation, is the rate-limiting step in creating a transgenic plant and these steps depend on a variety of tissue culture techniques. Callus is the most commonly used totipotent tissue from which whole plants can be regenerated. Callus is initiated by culturing transformed tissue such as leaf discs in a nutrient-rich medium containing growth hormones, and plantlets are formed on a medium containing regeneration hormones (cytokinins and auxins; Chapter 11). When both shoots and roots have formed, the plantlets are separated and carefully acclimatized for growth in a conventional agricultural or horticultural environment.

3. Applications of plant genetic engineering

Plant biotechnology (the combined application of plant tissue culture and molecular biological methods) is both a powerful research technique and a tool for the potential improvement of cultivars.

As a research technique, transformation provides an unparalleled means of studying the expression and function of plant genes from the level of the cell to the organism. Novel proteins can be engineered and the expression of specific proteins can be removed. In addition, transformation can be used to explore the functions of various sections of a gene or dissect the role of specific amino acid residues in a protein.

Producing new cultivars by traditional plant breeding is a slow and labour-intensive process. It takes several years to develop an improved variety

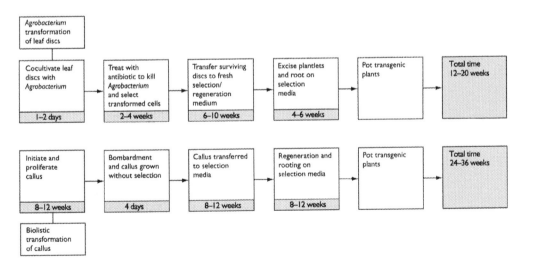

Figure 4.4

The key steps and the time scale involved for transformation and regeneration using Agrobacterium and microprojectile bombardment.

since several crosses must be made, seeds collected and germinated and the resulting plants grown before the results can be assessed. An additional and significant limitation of traditional plant breeding is that it can only use genes from within one species or closely related species for cultivar improvement. Furthermore, breeding typically involves transfers of whole sets of inherited characteristics and any undesirable traits introduced during initial crosses must later be meticulously bred out. Biotechnological procedures on the other hand offer the possibility of using genes from other kingdoms and species, improving the precision of transfer of desirable traits and generally speeding up the production of new cultivars.

The greater part of the current research in plant biotechnology focuses on food crops (80%); the remainder on non-food crops like cotton (*Gossypium* spp.), tobacco, ornamentals and on pharmaceutical production. Several crop plants have been genetically modified and some of these have been approved for commercial cultivation in some countries. Examples of genetically engineered crops that are currently grown commercially are described below.

3.1   Herbicide-resistant crops

In modern agriculture, weed-control is effected by selective herbicides that can suppress or kill yield-reducing weeds without harming the crop. The most common factors responsible for selectivity are: (i) the ability of the crop to metabolize the herbicide faster than the competing weed and (ii) the sensitivity of the protein targeted by the herbicide in the weed and the insensitivity of the corresponding protein in the crop plant. Several crops have been genetically modified to be specifically resistant to some herbicides (e.g. glyphosate, glufosinate and bromoxynil) that are widely used. Glyphosate (Roundup, *N*-(phosphonomethyl) glycine) is a non-selective herbicide, which means it is active against most plants. Glyphosate is not metabolized by plants and specifically inhibits the enzyme 5-enolpyruvylshikimate-3-phosphate synthase (EPSPS) that catalyses a critical step in the synthesis of aromatic amino acids. Crop plants have been engineered to be insensitive to glyphosate either by transferring a bacterial gene for EPSPS that is insensitive to glyphosate, thus providing a bypass of the plant's EPSPS, and/or a bacterial gene that encodes a protein that catalyses the metabolism of the herbicide.

3.2   Insect-resistant crops

Bacillus thuringiensis is a soil bacterium that produces insecticidal proteins known as the crystal proteins, Cry proteins or Bt toxins. There are different forms of protein produced by different subspecies of the bacterium, each of which is specifically active against one or two insect orders. When Bt protein is ingested by larvae of the target species, the toxin binds to cells of the gut causing lysis, paralysis and eventually death. Several crop plants, notably maize and cotton, have been transformed with various genes for Bt toxins to produce transgenic plants that are protected from insect attack and consequent yield losses.

3.3   Controlled ripening of fruit

Transgenic tomatoes have been engineered with delayed ripening of the tomato fruit. In one example, the genetic modification causes a down-regulation of the gene encoding polygalacturonase. Polygalacturonase is an enzyme that catalyses the cleavage of pectin chains in the cell wall of the fruit during ripening and which results in softening. By down-regulating polygalacturonase, tomato fruit soften more slowly during ripening than conventionally bred tomatoes. This allows harvesting to be delayed until the fruit has fully ripened, improving flavour and conferring improved processing properties.

3.4   Improving the nutritional value of food

Rice is a staple food in many parts of the world. The edible part of the rice grain, the endosperm, lacks provitamin A (β-carotene), which is converted in the human body to vitamin A. By engineering rice plants with a combination of three transgenes, the metabolic pathway for the biosynthesis of provitamin A in the endosperm was achieved (Potrykus, 2001). This provitamin rice known as Golden rice is still being developed.

Further reading

Chrispeels, M.J. and Sadava, D.E. (1994) *Plants, Genes and Agriculture*. Jones and Bartlett. Boston, London.

Chrispeels, M.J. and Sadava, D.E. (2003) Plants, genes and crop biotechnology. Jones and Bartlett, Boston, London.

Hansen, G. and Wright, M.S. (1999) Recent advances in the transformation of plants. *Trends Plant Sci.* 4: 226–231.

References

De Block, M., Herrera-Estrella, L., van Montagu, M., Schell, J. and Zambryski, P. (1984) Expression of foreign genes in regenerated plants and their progeny. *EMBO J.* 3: 1681–1689.

Potrykus, I. (2001) Golden rice and beyond. *Plant Physiol.* 125: 1157–1161.

Chapter 5

Culture facilities, sterile technique and media preparation

1. Introduction

Successful plant tissue culture requires laboratory facilities where aseptic conditions can be established and maintained. Aseptic technique is critical for plant tissue culture practice because the media and the environment in which plant materials are grown, are also ideal conditions for micro-organisms to proliferate. Fungal and bacterial contamination are amongst the hardest to deal with, as these organisms rapidly outgrow plant material and change the defined conditions of the growth medium resulting in poor growth or death of the plant culture. These deleterious effects are brought about by the contaminating organism(s) consuming the culture nutrients, excreting metabolites into the growth medium or colonizing the plant tissues. The basic aseptic techniques of handling and culturing plant materials were developed over many years and are designed to ensure that no contaminating micro-organisms enter the culture. Although not always essential, a carefully designed tissue culture unit is recommended to preserve sterile conditions and to achieve consistent results.

2. The basic laboratory layout and equipment

The prime purpose of a tissue culture laboratory is to enable the processing and culture of plant material in a sterile environment. To facilitate this purpose, certain basic facilities are required. These usually include the following:

■ a media preparation and sterilization area;
■ a sterile transfer area;
■ environmentally controlled incubators or a culture room.

In laboratories where plant transformation procedures will be routinely used, separate microbiological facilities may also be required for bacterial culture.

2.1 Media preparation area

The media preparation area should ideally be separate from the culture laboratory and should be equipped with the following:

■ storage space for chemicals, culture vessels and glassware required for the preparation of media;
■ refrigerator and freezer space to store stock solutions and chemicals;

- bench space with a balance, pH meter, water bath, hotplate and magnetic stirrers;
- a source of ultrapure water to be used for media preparation;
- an autoclave and a convection oven for the sterilization of media, glassware and instruments.

2.2 Sterile transfer area

In order to avoid any contamination that may come from open laboratory benches or from airflow, all manipulations involving culture transfers should be done in a sterile transfer area. Within this area, sources of electricity and gas should be available.

Plastic glove box

The simplest type of transfer area that can be used when relatively few transfers are required is a plastic glove box. This transfer unit can be sterilized by ultraviolet (UV) radiation and by spraying or wiping with 70% ethanol.

Laminar airflow cabinet

Most plant tissue culture facilities use a laminar flow cabinet (or hood) to provide an aseptic area for transfer work. These units are available commercially in different sizes and can be placed where required in the culture laboratory.

The laminar flow cabinet provides a constant flow of filter-sterilized air over the work surface. The air is passed first through a dust filter and then through a high-efficiency particulate air (HEPA) filter with a pore size of 0.2 μm, small enough to remove bacteria and fungal spores. The airflow is then directed either downward (a vertical flow system) or outward (a horizontal flow system) over a non-porous work surface. The constant flow of filter-sterilized air prevents non-sterilized air from entering the working area and thus creates a barrier against airborne contaminants (*Figure 5.1*).

The flow cabinet is usually illuminated by fluorescent light and may have a UV light for sterilization of the work surface. The UV light should be switched off when the cabinet is in use. The airflow should remain on continuously or should be run for at least 30 min before using. The work surface should be sterilized before and after use, by swabbing with 70% ethanol.

For safety reasons, it is advisable not to use a naked flame like a Bunsen burner in the cabinet as any fires (for instance involving ethanol) are rapidly intensified by the airflow and carried towards the operator. Instead, electrical sterilization devices like a bead or hot-air sterilizer may be used.

Placing of non-sterile materials in the hood should be avoided and as few things as possible should be stored in the cabinet. Any objects opened in the cabinet should be opened facing into the airflow. Before commencing work, hands and forearms should be carefully washed.

In order to minimize contamination, waste material and used apparatus should be removed immediately for sterilization and/or disposal.

Sterile transfer room

In laboratories where large numbers of cultures are routinely processed or large pieces of equipment are required, a useful arrangement would be a dust-free sterile transfer room. This room should have an overhead UV light

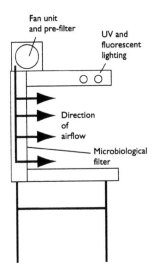

Fan unit
and pre-filter

UV and
fluorescent
lighting

Direction
of
airflow

Microbiological
filter

Figure 5.1

End elevation of a laminar flow cabinet. Arrows indicate the direction of the airflow in the cabinet. The best area to work is at the rear of the cabinet, with any open vessels facing into the airflow.

and a positive-pressure ventilation system equipped with a HEPA filter. It will need gas for a Bunsen burner and bench surfaces should be designed such that they can be easily cleaned and sterilized. The aseptic technique of standard microbiological practice should always be followed.

2.3 Culture room and plant growth facilities

The type of plant growth facilities required would depend on the quantity and type of plant material to be cultured. In general, plant tissue cultures are particularly sensitive to environmental conditions and should be grown under conditions where air circulation, temperature, humidity, light quality and photoperiod are well controlled.

The minimum culture requirement is a room with temperature control and lighting. If different plant cultures are grown in such a culture unit, it will not be practical to optimize conditions for all the types of culture and compromise settings may have to be adopted.

There is an optimal temperature for each type of plant culture, somewhere in the range 15–30°C. However, for most plants, the temperature is set at 25 ± 2°C. A major problem with a culture room is maintaining a constant temperature throughout the room as hot and cold spots (i.e. temperature gradients) may develop. In addition, the temperature in culture containers may be a few degrees higher than the room temperature because of the greenhouse effect. It is recommended that the room should be fitted with a safety alarm device to indicate when the temperature has reached pre-set maximum or minimum limits.

The culture room should be fitted with fluorescent lighting that can be adjusted for intensity and photoperiod. For mixotrophic cultures, light

intensities of approximately 30 μmol m^{-2} s^{-1} photosynthetic active radiation (PAR) are adequate. For autotrophic cultures, light intensities up to
250 μmol m^{-2} s^{-1} PAR will be required. In most cases, a regime of 16 h
light and 8 h dark is appropriate. A system where both light and temperature
can be programmed for a 24 h period would be advantageous.

Where possible, the culture room should have an adequate ventilation
system and the capability of controlling the relative humidity over a range of
20–98%. It should be noted that for cultures grown in standard culture containers, the headspace is usually saturated with water vapour. The relative
humidity in the container can be roughly controlled by cooling the base of
the container or adjusting the concentration of the agar or gel.

Many commercially available growth cabinets and purpose-built walk-
in growth rooms meet the above specifications. Clearly, an advantage of
individual cabinets is that the environmental conditions of each can be
optimized for growing a particular plant culture.

Where suspension cultures are to be grown, an illuminated orbital incubator with temperature control and lights can be used, or a shaking platform
with an orbital action can be placed in the culture room.

A simple plant tissue culture room may be set up as shown in *Figure 5.2*.
Culture containers are placed in translucent baskets and arranged on a
series of shelves or racks that are illuminated from above by fluorescent
tubes with a minimum gap of about 30 cm between lights and the shelf
beneath. Where cultures are to be grown in the dark, either light-proof
containers can be used or standard containers can be wrapped in aluminium foil. Electrical control gear that is likely to generate heat should
be placed outside the room. For reasons of electrical safety, such a culture

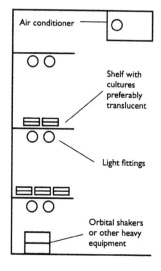

Figure 5.2

*End elevation of a typical plant growth room. Cultures in Petri dishes or jars are placed on
shelves under horticultural quality lighting. Orbital shaker platforms for suspension cultures
can be placed on the floor. Temperature and if necessary humidity can be maintained using
an air conditioner unit.*

arrangement is not suitable for growing plants in pots where watering is required.

In addition to the requirements of the culture room, glasshouse facilities may be needed for growing whole plants both before and after tissue culture manipulations. These facilities will also require control of temperature, lighting and relative humidity. Lighting for whole-plant growth rooms and glasshouses should ideally use horticultural tubes or bulbs to provide the appropriate light spectra.

Where genetically modified material is to be grown, rigorous containment of germplasm must be established and maintained in culture laboratories, culture rooms and glasshouses. It is important that the plant material is not allowed to escape to the environment unless licensed for release. In these circumstances, it must be possible to clean the growth area thoroughly (glasshouse, cabinet or room). Floors and benches should be impervious and crevices where material can accumulate should be avoided. Doors should be secured against casual access and space provided for separate waste management. Openings to the outside should be screened with a mesh fine enough to prevent movement of pollen and aphids. Local licensing regulations should be obeyed in detail in both the design of the facility and the material permitted to be grown.

3. Sterilization

One of the most important factors of plant tissue culture practice is establishing and maintaining aseptic conditions. Therefore, to avoid contamination and to preserve sterile conditions, tools, containers, media, explants and the working environment are sterilized and aseptic techniques are practised.

3.1 Sterilizing transfer areas and culture rooms

Transfer rooms can be sterilized by exposure to UV light. The time of exposure depends on the size of room. Sterilization should only be done when the room is not in use, as UV is harmful to plant material and damaging to eyes. Transfer rooms should also be sterilized by periodic (twice a month) spraying or wiping with commercial brands of bactericides and fungicides. Bench surfaces in transfer rooms should be sterilized before use by wiping with 70% ethanol. Laminar flow hoods are sterilized by wiping the work surface with 70% ethanol and running for 30 min. Culture rooms should be cleaned weekly with detergents and then wiped with 2% solution of sodium hypochlorite or 70% ethanol. Alternatively, floors, walls and work surfaces of culture rooms can be sterilized by treatment with commercial disinfectant such as Lysol.

3.2 Sterilizing instruments and culture vessels

Glass bead sterilization

Working tools such as metal instruments are easily sterilized by using a glass bead sterilizer that heats up to 275–350°C. The instruments are inserted into the heated glass beads for 10–60 s to destroy bacterial and fungal spores. The sterilized instruments are then placed in a sterile hood to cool until required.

Dry-heat sterilization

Glassware, metal instruments and materials such as aluminium foil can be sterilized by hot dry air at 130–180°C in a hot air oven for 2–4 h. All items should be sealed or wrapped in foil before sterilization. Please note that paper materials may decompose at higher temperatures.

Autoclaving

Glassware, filters, cotton wool plugs, plastic caps and instruments can be sterilized by autoclaving. Autoclaving is treatment by water vapour at high temperature and high pressure. The standard conditions used for sterilization are 121°C at 1.05 kg cm^{-2} for 15–20 min. Vessels should be closed with a cap or with aluminium foil and instruments should be wrapped in foil or wrapped in paper towels and placed in an autoclave bag. It is important that the steam can penetrate the items for sterilization. Autoclaves in different sizes are commercially available.

Glass pipettes should be plugged with cotton wool and autoclaved or baked in metal pipette cans or glass tubes sealed with cotton wool. Efficiency of autoclaving may be determined using autoclave type. Tips for automatic pipettes should be autoclaved in tip boxes. Instruments and pipettes should be dried in an oven at 80°C before use. Specialized items such as plastic mesh or coated glass slides may be sterilized by soaking overnight in 70% ethanol and drying in a sterile open Petri dish in a laminar flow cabinet for 20 min.

Flame sterilization

Frequently metal instruments previously sterilized by dry heat or auto-claving are removed from their wrapping and re-sterilized just prior to use by dipping in 95% ethanol and placing in the flame of a Bunsen burner. After use, instruments can be re-flamed before reusing. Safety is a major concern with this technique, because of the high flammability of ethanol and the risk of a flash fire. As described earlier, the danger is greater in a flow hood where there is a strong airflow directed towards the worker.

Pre-sterilized containers

A wide range of plastic containers suitable for plant tissue culture is now commercially available. These containers are sold already sterile (sterilized by radiation or ethylene oxide gas) and cannot be re-sterilized. Examples available from Sigma include the Phytacon™, a clear, plastic tub with a clip-fit lid and Phytatray™, a clear tray with a raised lid. Some are fitted with filters to allow gas exchange while maintaining sterile conditions (e.g. Magenta™).

3.3 Sterilizing explants

Preparing sterile explants is difficult because the tissue must be treated with disinfectants that are effective in destroying any microbial contamination without harming the explant tissue. The following procedural steps can be used to surface-sterilize explants.

(i) Explants are washed in a mild detergent. Leaf and herbaceous tissues may not require this step, but woody tissues and tubers must be washed thoroughly.

(ii) The washed tissue is thoroughly rinsed with running tap water for 10–30 min.
(iii) The tissue is then dipped for a few seconds (no longer than a minute) in 70% ethanol.
(iv) Under sterile conditions the explant is submerged in a disinfectant solution in a sealed bottle and gently agitated for 5–10 min. A wetting agent such as Tween 20 or 80 can be added to the disinfectant to improve contact between disinfectant and tissue. Disinfectants commonly used for plant tissue culture are shown in *Table 5.1*. For tissues that are difficult to decontaminate, vacuum infiltration of the disinfectant will enhance the sterilization by removing air pockets.
(v) At the end of the sterilization period, the disinfectant is decanted and the explants washed three or more times in sterile distilled water, gently blotted with sterile paper towels and then used to initiate the tissue culture.

For tissues that are difficult to disinfect, it may be necessary to repeat the sterilization procedure after 24–48 h, before selecting the final explant. This allows any remaining microbes time to develop to a stage at which they are more easily destroyed by the disinfectant.

Other methods to reduce high contamination of tissue include taking explants from stock plants grown in a greenhouse or growth chambers rather than from plants grown in the field and treating the explant with antibiotics and fungicides. The latter compounds are not always effective and are often phytotoxic.

3.4 Sterilizing culture media

Two methods are used to sterilize culture media and culture media supplements:

- autoclaving for thermostable ingredients;
- filter sterilization for thermolabile compounds.

Table 5.1. Disinfectants commonly used to sterilize explant material for plant tissue culture

Disinfectant	Concentration (%)	Comments
Sodium hypochlorite	0.5–5	Activity based on available chlorine and oxidizing action
Commercial bleach	10–20	Contains approximately 5% sodium hypochlorite
Calcium hypochlorite	9–10	A solid that must be dissolved and filtered.
Hydrogen peroxide	3–12	Activity based on strong oxidizing action
Benzalkonium chloride	0.01–0.1	Cationic surfactant. Permeates biological membranes
Mercuric chloride	0.1–1.0	Denatures proteins. Very toxic, requires special disposal for waste solutions
Ethanol	70–95	Denatures proteins

Basal media, ultrapure water and any other heat stable components are placed in glass vessels that are sealed with plastic caps (loosely fitted), aluminium foil or cotton wool plugs and autoclaved at 121°C at 1.05 kg cm^{-2} (100 kPa) for 15–20 min. Large volumes of 2–4 l require a longer time, 30–40 min. Pressures should not exceed 1.05 kg cm^{-2} (100 kPa) as higher pressure can cause the breakdown of carbohydrates in the medium.

Solutions of media components such as vitamins, amino acids and hormones that are thermolabile are filter-sterilized. Sterile filters (e.g. Millipore filters) with a porosity no larger than 0.22 μm are used and the sterile filtrates are collected in sterile containers.

4. Media preparation

The components of plant growth media have been described in Chapter 2. There is a vast range of media commercially available, which are derived from early experimental work in plant tissue culture. These include the media of Gamborg, Hoagland, Litvay, and Schenk and Hildebrandt. There are also over 20 variants based on Murashige and Skoog basic medium embracing a large number of speciality cultures. Generally, all these media are described as containing the complete range of micro- and macro-nutrients necessary for plant tissue or cell culture. Others have a reduced component mix to allow the experimenter to vary medium composition as required. Media either can be bought commercially as prepared powder mixtures or can be made up from basic components. In the latter case, great care must be exercised to ensure consistency from batch to batch. If large amounts are being prepared, all solid ingredients need to be finely ground either with a pestle and mortar or with a mill, to ensure homogeneity. Powdered salt mixtures are extremely hygroscopic and should be protected from atmospheric moisture by storing in a desiccator over a drying agent. Alternatively, it may be convenient to prepare inorganic macronutrients and inorganic micronutrients as separate stock solutions (*Table 5.2*).

A major component of some culture media, not normally included in the basal salt mixtures, is the carbon source. This is usually a sugar that can be easily metabolized, like sucrose. Sucrose is normally added at a concentration of 1–3% w/v (10–30 g l^{-1}).

The culture medium may also require single vitamin additions or more commonly, vitamin mixes. Commonly used vitamins and other organic compounds are listed in *Table 5.2*. These are sold as concentrated, filter-sterilized solutions. It is important to choose those that have been previously tested in plant tissue culture. Vitamin mixes such as Gamborg's or Litvay's are also available as powders to be dissolved as 1000× stock solutions. The solutions should be filter-sterilized, aliquoted into 5- or 10-ml sterile tubes and stored at −20°C for subsequent use.

Many plant cultures require growth regulators (hormones). The most frequently used plant hormones are sold as powders to be dissolved, as described in *Table 5.3*. It is useful to prepare a stock solution that is 1000× the final concentration to facilitate small volume additions. Hormones are added before autoclaving or filter-sterilized and added after autoclaving (*Table 5.3*).

Table 5.2. Composition of some other basal plant tissue and cell culture media

Compound	Concentration in final medium (mg l^{-1})				
	White's	Schenk and Hildebrandt's	Litvay's	Gamborg B$_5$	Chu N6
Macronutrients	–	–			
$(NH_4)_2SO_4$	–	–	–	134	463
KH_2PO_4	12	–	–	–	400
KNO_3	81	2500	2500	2500	2830
$CaCl_2$	–	151	151	–	125.33
$Ca(NO_3)_2$	200	–	–	–	
$CaCl_2.2H_2O$	–	200	–	150	–
$MgSO_4$	360	195.4	195.4	122.09	–
$MgSO_4.7H_2O$	–	–	–	–	90.37
NH_4NO_3	–	–	–	–	–
Micronutrients					
$CoCl_2.6H_2O$	–	0.1	0.1	0.025	–
$CuSO.5H_2O$	0.01	0.2	0.2	0.025	–
$CuSO_4$	–	–	0	–	–
$Fe(SO_4)_3$	2.46	–	–	–	–
H_3BO_3	1.5	5	5	3	1.6
KCl	65	–	–	–	–
KI	0.75	1	1	0.75	0.8
$NaH_2PO_4.H_2O$	16.5	–	1.25	130.5	–
$NH_4H_2PO_4$	–	300	–	–	–
$MnSO_4.H_2O$	5.04	10	10	10	3.33
$Na_2MoO_4.2H_2O$	–	0.1	0.1	0.25	–
$ZnSO_4.7H_2O$	2.67	1	1	2	1.5
Molybdenum trioxide	0.001	–	–	–	–
Na_2SO_4	200	–	–	–	–
Iron source					
Na_2EDTA	–	20	20	37.3	37.25
$FeSO_4.7H_2O$	–	15	15	27.8	27.85
Organic compounds					
Myo-inositol	–	1000	–	1000	–
Nicotinic acid	–	5	–	1	
Thiamine-HCl	–	5	–	10	
Pyridoxine-HCl	–	0.5	–	1	
Sequestrene 330 Fe	–	–	–	28	–
Yeast extract	100	–			
Sucrose	20	30			30

It is possible that the medium will require an undefined supplement such as coconut water, casein hydrolysate or banana powder, all of which can be autoclaved. Coconut water, used at 1–10% v/v, is extracted from coconuts (see *Protocol 5.4*). Banana powder is available as lyophilisate and is used at 25–40 g l^{-1}. Casein hydrolysate may be used at 100 mg l^{-1}. Amino acid supplements such as glutamine are very heat-labile and need to be

Table 5.3. Hormone additions to basal media

	Compound	Molar mass	Concentration range in media	Solution preparation
Auxins	Indole-3-acetic acid (IAA)[a]	175.2	$10^{-7}-10^{-5}$	Titrate into aqueous solution with KOH to make a 10 mM stock.
	2,4-Dichlorophenoxyacetic acid (2,4-D)	221.0		
	3-Indolebutyric acid (IBA)	203.2		
	1-Naphthaleneacetic acid (NAA)	186.2		
Cytokinins	Zeatin (Zea)	219.2	$10^{-7}-10^{-5}$	Dissolve in dilute KOH or aqueous ethanol to produce a 10 mM solution.
	6-Furfurylaminopurine (Kinetin)	215.2		
	6-Benzylaminopurine (BAP)	225.2		
	N-isopentenylaminopurine (2iP)	203.3		
Gibberellins	Gibberellic acid (GA$_3$)	346.4	$10^{-7}-5 \times 10^{-6}$	Soluble in water. Make a 10 mM stock solution

Note that solutions containing cytokinins and gibberellins should not be autoclaved. They are usually sterilized by passage through a 0.22 µm filter.
[a]IAA is oxidized readily by plant material.

filter-sterilized. Antibiotics or antimycotics should only be added after auto-claving. Again, these are available as plant tissue culture tested and need to be dissolved to a suitable concentration, filter-sterilized and can be stored frozen. Most antibiotics will be stable at –20°C for 3 months; however, Rifampicin should be freshly prepared and Ampicillin is extremely light sensitive.

If solid medium is required, a gelling agent is added before autoclaving. There is a variety of gelling agents available, as described in Chapter 2. Agar, the most commonly used gelling agent, is available in a range of purities, but for plant tissue cultures, the most pure should be used. Agar is used at a concentration of 0.8 and 2% w/v.

All media should be adjusted to the appropriate pH before autoclaving after all additions (with the exception of agar) have been made. As a general rule, hormones dissolved in alcohol do not affect pH, while other additions are made in such small volumes that the effect is negligible.

Prior to autoclaving, flasks containing media should be loosely sealed, for instance with plastic caps, aluminium foil caps, or with cotton or foam bungs. For safety reasons, a facemask and thick, waterproof gloves should be worn while unloading the autoclave. Although most autoclaves are fitted with safety locks, at the end of the run it is essential to ensure that the temperature is safely below boiling point and that the pressure has returned to atmospheric pressure before opening the autoclave. Failure to do this can result in sudden and spontaneous boiling of the media, which can cause serious burns.

It is useful to have a clean area (laminar flow cabinet or similar) in which to allow media to cool after autoclaving and to allow solid media to set. Containers should be level and not disturbed until media have fully cooled. Plates, vessels and liquid media can then be stored for a few days in a refrigerator until required. Solid media may not be frozen. It is very important that packets of plates and containers are fully labelled with contents, date and experimenter's initials at the time of pouring. It is especially easy to confuse plates of different media prepared at the same time.

5.   Contamination

Cultures should be checked 3–5 days after initiating or subculturing for contamination. Contamination can come from several organisms.

■   Bacteria are the most frequent contaminants. Bacterial contamination is commonly introduced with the explant and is characterized by a slimy appearance. Bacterial contamination can be many colours such as white, cream, pink or yellow.

■   Fungi may enter cultures on tissue explants or from airborne spores. Fungal contaminations are recognized by their 'fuzzy' appearance and occur in many colours.

■   Yeasts are frequent contaminants that may enter cultures on the explant or may be present in the air.

■   Viruses and mycoplasma-like organisms can be sources of contamination and are not easily detected.

■   Insects, such as ants, thrips and mites can be troublesome contaminants in cultures. Thrips emerge from eggs present in the explant and contamination of mites is usually an indicator of poor laboratory hygiene. Thrips and mites are responsible for spreading fungal spores and are often detected through the appearance of fungal growth along the paths of the insects.

Contamination can be introduced in several ways:

■   introduced contamination – resulting from poor laboratory hygiene and/or aseptic technique;

■   initial contamination – due to incomplete sterilization of the explant;

■   latent contamination – usually resulting from endogenous bacteria present in the tissue explant that grow and multiply long after culture initiation. Latent bacteria are a troublesome problem as they may be transferred easily at the time of subculturing.

Positive identification of the contaminating organisms can be established by visual screening or by the process known as indexing. This involves taking a culture sample and inoculating media that are specific for bacteria, fungi or yeasts. The media frequently used are nutrient broth (NB) containing salts, yeast extract and glucose to stimulate bacterial growth, or potato dextrose agar (PDA), for the growth of fungi and yeasts. However, unless cultures are particularly valuable, contaminated material should be disposed of without delay by autoclaving to prevent it spreading to other cultures.

6.   Disposing of contaminated waste

Used agar plates should be placed in autoclave bags and the bags placed inside a plate box or metal bucket, designed to withstand the pressures of autoclaving. This arrangement prevents spillage of agar during autoclaving. Used empty Petri dishes, tips, plant material and disposable pipettes may be autoclaved in a bag if local regulations permit. Autoclaved material may be

Table 5.4. Autoclave conditions for disposing of contaminated waste

Item	Temperature	Time
Plastic waste (Petri dishes, plate boxes, no glass)	134°C	10 min
Mixed load plastics and fluids in glass or metal buckets	121°C	35 min
Laboratory coats	134°C	5 min
Fluids in large volumes (5–10 l)	121°C	At least 30 min

then disposed of as for normal waste. Contaminated liquids may be either autoclaved or treated with freshly prepared sodium hypochlorite (5% final concentration) overnight. These liquids may be disposed of to drains if local regulations permit. Used sharps and needles should be disposed in the correct hazard bins, destined for burning. Used tools such as forceps may be either autoclaved or heated in a bead sterilizer. Suitable autoclave conditions for disposing of contaminated waste are presented in *Table 5.4*.

7. Safety in the laboratory

Autoclaves operate under high temperatures and pressures, and should be used with great care. Most autoclaves have pressure and temperature locks that will not allow the operator to open the machine until it is completely safe to do so. Even when the autoclave has cooled sufficiently to allow opening, the media contained in the bottles may still be very hot and can boil over and spill with the risk of serious burns. For these reasons, insulated, waterproof gloves and a clear full-face visor should be worn when unloading an autoclave. The gloves need to be waterproof because the outside of the bottles or vessels will be very wet and the water will track through a non-waterproofed material, conducting heat through to the operator. The most commonly occurring accidents related to autoclaves are those that involve sealed units, particularly glass bottles. Tightening the lids of screw-cap bottles prior to autoclaving can cause them to explode. Therefore, each bottle must be checked to ensure that the cap is loose enough to allow steam to escape before it is loaded into the autoclave. Autoclaved media in glass bottles should be handled using thick gloves until it has cooled sufficiently to pour. Similar precautions should be taken removing articles from ovens. Another risk in the tissue culture laboratory is that of hazardous chemicals. Care must be taken in the handling and disposal of these according to the COSSH regulations (in the UK) or other appropriate local regulations.

It is important to pay careful attention to the local regulations governing the control of work involving genetic modification (GM). All plant material or equipment that has been in contact with GM material must be autoclaved before it leaves the laboratory. In addition, transformed plant material should only be grown in areas that have pollen and aphid screens on all windows or vents and a double door arrangement to act as a pollen trap. Glasshouse facilities should be divided to separate GM and non-GM plants so that there is no possibility of cross-pollination. Waste materials should be disposed of according to local regulations; living material and waste media should be autoclaved before disposal.

Protocol 5.1

Hygiene in the tissue culture laboratory

The tissue culture unit should be designed for easy cleaning. Bench surfaces should be laminated with integrated sinks moulded into them and under-bench cupboards should be raised from the floor. The floor and skirting should be easy to clean without joints or crevices. Bench-top shelves, up-stands and electrical fittings should be moulded without cracks or gaps. All surfaces should be cleaned regularly with a proprietary disinfectant containing sodium hypochlorite. Where practical, heating and ventilation should cause minimal air-movement, especially near the laminar flow cabinets where conflicting airflow can result in material being contaminated.

Equipment

Laboratory coat reserved for use in tissue culture facility

Laboratory paper towels

Laminar flow cabinet situated in a clean room

Autoclavable waste boxes for Petri dishes

Plastic waste bins lined with autoclave bags for dry plastic waste (e.g. pipettes or tips)

Jugs containing either fresh hypochlorite or other sterilant for contaminated liquids.

Materials and reagents

Spray containing 70% ethanol

Hand wash sink with liquid soap dispenser and elbow taps.

Protocol

1. Leave all personal possessions and coats in a secure area outside the tissue culture laboratory.

2. Tie back hair and wash hands.

3. Put on clean tissue culture lab coat.

4. Thoroughly wipe down laminar flow cabinet with 70% ethanol. Start fan and leave running for 30 min.

5. Meanwhile, spray equipment to be used with alcohol and place in flow cabinet. Ensure that marker pen used to label plates or equipment is not alcohol soluble. If it is, only wipe unmarked parts of plates with tissue soaked in alcohol.

6. The most sterile zone is at the back of the cabinet so use this area.

7.   Open media bottles or Petri dishes into the air flow.

8.   Discard unwanted equipment or media as soon as convenient. Maintaining sterile conditions is easier when the working area is uncluttered. Autoclave bags for waste should be placed adjacent to the cabinet.

Safety note

Take care when using 70% alcohol spray. Alcohol is toxic and flammable; avoid skin contact, inhalation or ingestion. Keep well away from naked flames or sources of ignition.

Protocol 5.2

reparing basal medium from commercial formulations or from individual components

The required basal medium can be prepared either by dissolving the commercially available basal salts mix or by using a table of components (examples are given in *Table 5.2*) and preparing the mix from the individual ingredients. An alternative is to prepare macronutrient and micronutrient stock solutions (*Table 5.2*) and store these at 4°C until used. Organic compounds should not be stored in solution for more than 2 weeks. Heat-labile compounds such as plant growth regulators, antibiotics or glutamine should be filter-sterilized (see *Protocol 5.3*) and added when the medium has been autoclaved and cooled.

Equipment

Balance

Beaker that is at least 50% larger than the volume of medium to be prepared

Spatula

Magnetic stirrer and bar

Measuring cylinder (1 l)

Glass or plastic bottles for storage (N.B. Glass vessels should not be stored at −20°C)

pH meter

Vessels for autoclaving (conical flasks for liquid media or Duran bottles for solid media)

Autoclave and autoclave tape

Sterile containers to be used for solid media immediately after autoclaving (e.g. Petri dishes or Phytatrays™)

Automatic pipette (1 ml) and tips

Laminar flow cabinet.

Materials and reagents

Ultrapure water

Either a commercially prepared basal salts medium (e.g. one of the media shown in *Table 5.3*) or macronutrient and micronutrient stocks (stored at 4°C) and iron stock (*Table 5.2*).

Organic compounds stock (stored at −20°C)

Sucrose

Gelling agent (e.g. Agar; see section 5 above).

1 M NaOH or 1 M HCl to adjust pH.

Protocol

1. Add approximately half the final volume of ultrapure water to the beaker and position it on the magnetic stirrer.

2. Start the magnetic bar spinning.

3. Weigh out the appropriate amount of basal salts or measure the appropriate volume of stock solutions. Add to the water stirring in the beaker.

4. Weigh out the sucrose and add to the beaker.

5. Measure and add any other heat-stable components, such as undefined media supplements. Details of such additions are presented with specific protocols in subsequent chapters.

6. When all solids have dissolved, adjust to final pH, using 0.1 M NaOH or HCl as necessary. Sensitive adjustments may be made using an automatic pipette.

7. Make up to final volume using a measuring cylinder.

8. If preparing liquid medium, measure out into conical flasks (30 ml in a 100-ml flask, or 75–100 ml in a 250-ml flask). The vessels can be capped with aluminium foil sheets, ensuring that no opening is left through which contaminants may enter after autoclaving. Some laboratories also seal flasks with either non-absorbent cotton wool or foam bungs.

9. If preparing solid or semi-solid media, weigh agar (commonly between 8 and 10 g l^{-1}) and add directly to the vessel to be autoclaved. Pour the medium on to the agar, washing the entire agar down to the bottom the flask. Do not over-fill the bottle; for example, 400 ml in a 500-ml bottle is the maximum volume recommended.

10. Solid or semi-solid media can be autoclaved either in the end-use vessel or in a screw-cap bottle. After autoclaving, the media can be cooled and poured into pre-sterilized plastic containers or glass containers sterilized by autoclaving. Few plastics can be autoclaved. If in doubt, check by autoclaving one piece in a glass beaker.

11. Transfer vessels to autoclave. Media vessels sealed with foil can be safely autoclaved but screw-top bottles should be autoclaved with the caps loose. They should be tightened as soon as possible after the autoclave cycle is complete. Check that everything needed (such as pipette tips, fresh foil caps or forceps) are included.

12. Autoclave at 121°C (100 kPa, 15 p.s.i.) for 15 min. This may be reduced for vulnerable media such as those rich in sugars. Browning of the medium is a warning sign. The lowest parameters for sterilization are 118°C (70 kPa, 10 p.s.i.) for 10 min. An alternative to autoclaving is to filter-sterilize the most heat-labile component(s) and add them post-autoclaving (see *Protocol 5.3*).

13. After the autoclave cycle has been completed and the chamber has cooled to at least 70°C, the contents may be removed. Tools or tips may need to be dried in an oven prior to use. Media already in end-use vessels should be left, preferably, in a flow cabinet, to cool and (if solid) set. Ensure the vessels are level. If media is to be transferred to a sterile plastic container, it should be

allowed to cool until hand hot, when sterile antibiotics or vitamins may be added if necessary, and poured in a laminar flow cabinet.

14. It is important to check that the labels on the media are intact. Petri dishes of media can be stored in the original plastic tubing, labelled with the composition and the date.

15. Media are best stored at 4°C, never frozen and should not be used when more than 2 weeks old.

Note

Hormones and other additions are frequently added to basal media; see *Protocol 5.3* below.

Safety note

Tightening the lids of Duran or similar screw-cap bottles prior to autoclaving is dangerous and can lead to them exploding in the autoclave. They should be left resting on the bottle top until the autoclave has been opened and the contents have cooled to less than 70°C. To avoid contamination, the bottles should be transferred to a laminar flow cabinet immediately after they are removed from the autoclave.

Protocol 5.3

Preparation of hormone additions to basal media

Equipment

Syringe filters (0.22 μM)

10-ml measuring cylinder

Fine balance

Luer-lock syringes (10 ml)

Sterile containers for stock solutions (plastic Universal bottles)

pH meter with fine probe

Pipette

Magnetic stirrer and small stirrer bars.

Protocol

Prepare individual stock solutions for each of the hormones needed in plastic Universal tubes. *Table 5.3* shows the molar mass of the major types of hormones used in plant cell culture. It is suggested that a 10 mM stock solution is appropriate for most applications. The example of preparing stock solutions of 1-naphthaleneacetic acid (NAA) and 6-benzylaminopurine (BAP) will be given.

1. Weigh out 18.62 mg of NAA into a Universal bottle.

2. Add 7 ml of distilled water with stirring (magnetic stirrer).

3. Bring to pH 6.5 with constant stirring by adding 0.5 M KOH dropwise.

4. Transfer to a 10-ml measuring cylinder, bring to volume, mix again.

5. Transfer to a Universal bottle for storage.

6. Weigh out 22.52 mg of BAP into a second Universal bottle.

7. Add 7 ml of 10% aqueous ethanol with stirring (magnetic stirrer).

8. When fully dissolved, transfer to a 10-ml volumetric flask and bring to volume with ultrapure water.

9. Filter both solutions through 0.22 μm syringe filters into sterile plastic Universal bottles under sterile conditions in a laminar flow cabinet.

10. Use immediately or store at −20°C.

11. To use after freezing, thaw and take appropriate aliquot aseptically before adding to medium.

12. Loosen the tops of both the sterile growth regulator bottle and the container of freshly autoclaved medium.

13. Push the automatic pipettor into the top of a sterile tip presented in a tip box and lift the tip out without touching the sides.

14. Using the free hand, lift the tip off the bottle containing the additive and withdraw the set amount.

15. Replace the lid loosely on the bottle and use the same hand to remove the top of the bottle of medium.

16. Dispense the additive into the medium being very careful not to touch anything.

17. Discard the tip into a waste container.

18. Cap the medium bottle and mix thoroughly by rotating horizontally on the bench. It is not the best sterile practise to invert the bottle.

19. Liquid medium can be stored or used while solid medium can be poured and allowed to set.

Protocol 5.4

Preparation of coconut water supplement for basal media

Equipment

Hand or electric drill with 8-mm wood bit

Bench vice

Funnel (500-ml)

Filter paper (Whatman number 1)

Retort stand, clamps and bosses

Conical flasks

Plastic bottles for storage (100–200-ml).

Protocol

1. Purchase a sack of good-quality fresh ripe coconuts from a wholesaler.

2. Mount a coconut in a bench vice and drill two 8-mm holes through the shell.

3. Drain the coconut water (liquid endosperm) via the filter funnel and Whatman number 1 filter paper into a conical flask.

4. Check the product of each coconut by eye and smell for quality. A yellow colour or an acid smell is a contraindicator.

5. Pool the coconut water from the entire batch and mix.

6. Store frozen at $-20°C$.

Chapter 6

Callus cultures

I. Introduction

Callus is an amorphous mass of unorganized thin-walled parenchyma cells. When a plant is wounded, callus formation occurs at the cut surfaces and is thought to be a protective response by the plant to seal off damaged tissues. The formation of wound callus has been observed in almost all groups of living plants.

In culture, callus is initiated by placing a fragment of plant tissue (an explant) on solid culture media under aseptic conditions. Callus is induced and formed from proliferating cells at the cut surface of the explant tissue. Depending on the species, callus can be initiated from a variety of tissues by employing the appropriate growth medium. However, rapid cell division can be more easily induced in some tissue than in others. The *in vitro* formation and proliferation of callus is enhanced by the presence in the medium of hormones (auxins and cytokinins, see Chapter 2) that promotes cell division and elongation.

2. Origin of callus

During the *in vitro* initiation of callus, the cell differentiation and specialization that occurred in the parent plant is reversed and cells of the explant become dedifferentiated. The process of dedifferentiation is characterized by changes in metabolic activity, the disappearance of storage products and rapid cell division that gives rise to undifferentiated and unorganized parenchyma cells (see Chapter 2). The lack of structural organization persists as the callus grows, although a homogeneous callus consisting entirely of parenchyma cells is rarely observed. As growth of the callus proceeds, centres of meristematic activity are formed. Random and rudimentary cambial zones may give rise to regions showing vascular differentiation in the form of sieve elements, suberized cells or tracheary elements. Some of the centres of meristematic activity form nodules that may be precursors of shoot apices, root primordia or incipient embryos, which are capable of further development if exposed to suitable culture media.

3. Types of callus

Callus varies widely in its general appearance and in other physical features. The variation depends on the parent tissue, the age of the callus and the growth conditions. Callus may be white, green or highly coloured due to the presence of anthocyanin pigments. Callus may consist of loosely packed cells and be friable (i.e. easily crumbled or fragmented, *Figure 6.1*), or may be lignified, with densely packed cells and hard in texture (non-friable).

Furthermore, callus may or may not be embryogenic, that is, able to form embryos either spontaneously or when grown under suitable conditions. In monocotyledonous grasses like maize, these features have been used to define two types of callus. Type I callus is non-friable, regenerates somatic embryos and organs and frequently produces leaf-like structures. Type II callus is friable, undifferentiated and regenerates only somatic embryos.

Callus cultures are also predisposed to genetic instability and as a result, variation in phenotype within the same culture may occur. This phenomenon known as somaclonal variation is dealt with in greater detail in Chapter 12.

4. Role of callus in embryogenesis, organogenesis and cell culture

Formation of callus is a fundamental step in the *in vitro* culture of many types of plant cells and tissues and in some methods of plant genetic manipulation (*Figure 6.2*). For example, *in vitro* callus provides the most frequently used totipotent cells from which whole plants are regenerated via either organogenesis or somatic embryogenesis (*Figures 6.3* and *6.4*). Callus is often used as the target tissue for genetic transformation and callus formation is initiated for the regeneration of plants after transformation of other tissues. Dispersal of friable callus into single cells or clumps of cells is used universally as the method of initiating cell suspension cultures.

5. Initiation and establishment of callus cultures

The successful establishment of a callus culture requires consideration of four important areas:

- selection of suitable parent material;
- choice of explant and method of isolation;
- the media and culture conditions required;
- optimization of the culture conditions.

Figure 6.1

Cells of a friable callus of Nicotiana *suitable for breaking up to form suspension cultures, viewed by dissecting microscope.*

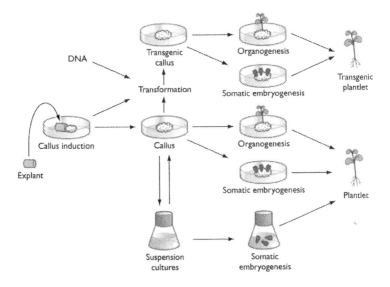

Figure 6.2

Alternative applications of callus in plant tissue culture and genetic transformation.

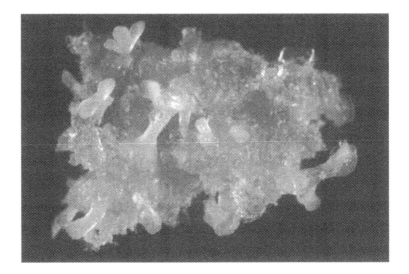

Figure 6.3

Shoots developing from callus formed on leaf squares of Nicotiana *grown in the presence of IBA and BAP.*

Selection of suitable starting material requires seedlings or plants that are growing vigorously and are free of disease. If the callus culture is to be used as part of a micropropagation programme, the parent plant must show the desired genetic characteristics. Such a plant is known as an elite mother plant.

Explants are isolated from elite material, in sterile conditions. The mother plant material and explants are decontaminated and surface sterilized as

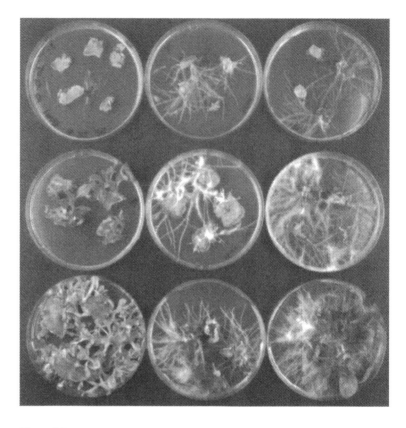

Figure 6.4

Nine plates from a 'growth square' of tobacco leaf material grown on media containing varying auxin and cytokinin concentrations showing the formation of roots, shoots and leaves. The plates shown are (top row) plates 1–3, (second row) plates 6–8 and (third row) plates 9–11 of the growth square described in Table 6.1.

described in Chapter 5. Explants are removed with sterile instruments. The method for removing the explant is unimportant, though the act of wounding is important in producing a wound response. Leaf pieces can be cut using sterile scissors, a scalpel or a cork-borer. Stem and storage root cores are usually taken with a cork borer and then cut transversely with a sterile scalpel. Explants taken from internal plant tissues, for instance, cores taken from within the stem (pith) or from taproots, may only require brief surface sterilization after removal from the plant. In each case, care must be taken to keep the explant sterile throughout and to discard any plant material that may be contaminated.

The formation of callus is usually achieved by placing the sterile explant onto an appropriate solid growth medium in a container such as a Petri dish or culture tube. The container is closed and incubated under the prescribed environmental conditions. After a period of 2–3 weeks, callus formation is observed first as outgrowths at the edges of the wounded material, and later as a mass of cells which gradually grows over the original plant material. Callus formation may be preceded by expansion growth of the plant tissue.

Table 6.1. Growth 'square' to test media concentrations of auxin and cytokinin for optimal callus growth

IAA (μM)	Kinetin (μM)				
	0	0.5	2.5	5	10
0	1	2	3	4	5
0.5	6	7	8	9	10
2.5	11	12	13	14	15
5	16	17	18	19	20
10	21	22	23	24	25

Optimization of the growth of the callus may require initial trials with several different media; transfer of established callus to new media with varied compositions, or selection of light-coloured (no browning) friable fragments for subculture. Examples of media for the optimal growth of callus from several species are given in the protocols of this chapter. If callus production from other species is required, it is advisable first to undertake a literature search to see whether media have previously been optimized for that species. If no information can be found, use one of the standard media described in Chapter 5 and set up a number of agar plates with varying auxin and cytokinin concentrations. One of the best ways to do this is to set up a growth (Latin) 'square' of 25 plates, with five auxin concentrations in combination with five cytokinin concentrations (see Table 6.1 and Figure 6.4). For this growth trial, the inocula of callus should be of uniform size and large enough (about 20 mg) to ensure growth on a new medium.

Protocols are presented in this chapter for callus production from *Arabidopsis thaliana* seedlings (based on the methods described by Blackhall, 2001), from imbibed seeds of gymnosperms and angiosperms (based on Attree *et al.*, 1996 and our own laboratory and Sellmer *et al.*, 1996) and from roots and shoots of monocots and dicots, using methods practised in our laboratory.

6. Monitoring the growth of callus

The growth of callus is monitored by measuring the fresh weight and dry weight of the cultures. For fresh weight measurements, the callus is transferred to a pre-weighed foil weighing boat and rapidly weighed. It is important that weighing of the callus is done as quickly as possible after removal from the container as the culture loses water rapidly in open air. After the fresh weight measurement, the callus and boat is then oven dried (60°C) overnight and re-weighed to obtain a dry weight. As these measurements result in the death of the culture, they are therefore only useful for optimizing growth conditions. Rough estimates of growth may be made by estimating the diameter of the callus *in situ* and the health of the culture can be assessed by observing its colour. When the medium is unsuitable, growth is slow and browning of the culture occurs. Usually, rapid growth and a light coloured callus indicate a healthy culture. A friable callus of crumbly appearance is very suitable for breaking up, either for subculturing or to produce a suspension culture. In our experience, darkening of the callus is often associated with cell death and it is therefore best to avoid such material. Sometimes discoloration of the culture is an indication of microbial contamination, in which case the culture should be disposed of by autoclaving.

Callus cultures are normally maintained at around 22–25°C under low-intensity fluorescent light with a dark/light cycle of 8 h:16 h. Callus cultures can be maintained under these conditions for several years with sub-culturing every 3–6 weeks depending on the species and the growth rate of the culture. Callus cultures are often used as a means of preserving germplasm by storage and maintenance under slow-growth conditions (Chapter 3).

7. Genetic transformation of callus

Callus cultures are important elements of almost all current practical transformation strategies (see Chapter 4). Callus cells are frequently used as targets for transformation both by microprojectile bombardment and by *Agrobacterium tumefaciens*. In addition, when other material such as leaves or embryos are used as targets for transformation, callus is initiated either before transformation or during the recovery of transformed cells. After transformation, a callus stage usually precedes the differentiation of roots, shoots or embryos during the induction of plant regeneration.

Agrobacterium-mediated transformation of dicots utilizes sterile explants of plant material such as leaves, cotyledons, stems, germinating seeds and callus cultures. Typically, leaf discs or pieces are used as the material for transformation and may be transformed without or with prior induction of callus. Leaf pieces are incubated with the bacteria for 5–10 min to allow the bacteria to infiltrate the tissue, then blotted dry and transferred to a fresh plate of solid medium where the leaf tissue and bacteria are co-cultivated for 2 days. By the end of this period, the bacteria are seen to overgrow the leaf surface. The leaf discs are washed clear of bacteria and then transferred to a medium containing antibiotics (e.g. carbenicillin and timentin) to kill the *Agrobacterium* and antibiotics (e.g. hygromycin) to select transformed cells. Discs are transferred at weekly intervals to new media and by adjusting the hormonal composition of the media, transgenic callus and/or plantlets can be induced to form.

Monocot callus (e.g. maize) derived from embryogenic tissue is the commonly used material for direct transformation by microprojectile bombardment. A method for the direct transformation of maize callus is presented in *Protocol 6.6*, and is based on the published procedure of Fromm (1996).

References

Attree, S.M., Rennie, P.J. and Fowke, L.C. (1996) Induction of somatic embryogenesis in conifers. In: *Plant Tissue Culture and Laboratory Exercises* (eds. R.J. Trigiano and R.J. Gray). CRC Press, Boca Raton, Florida.

Blackhall, N. (2001) Arabidopsis, the compleat guide – http://www.arabidopsis.org/comguide/chap_2_tissue_culture/13_cell_suspension_culture.html

Chu, C.C., Wang, C.C., Sun, C.S., Hus, C., Yin, K.C. and Chu, C.Y. (1975) Establishment of an efficient medium for anther culture of rice through comparative experiments on the nitrogen sources. *Scientia Sin.* **18**: 659–668.

Fromm, M. (1996) Production of transgenic maize plants via microprojectile-mediated gene transfer. In: *The Maize Handbook* (eds. M. Freeling and V. Walbot). Springer, New York, pp. 677–684.

Sellmer, J.C., Ritchie, S.W., Kim, I.S. and Hodges, T.K. (1996) Initiation, maintenance and plant regeneration of type II callus and suspension cells. In: *The Maize Handbook* (eds. M. Freeling and V. Walbot). Springer, New York, pp. 671–677.

Protocol 6.1

Preparation of plant material and explants

Equipment

Sterile:

Conical flask (250 ml), sealed with aluminium foil

Petri dish

Tea strainer or muslin squares, autoclaved in foil packets

Sterile containers for sterilizing plant material (e.g. beaker sealed with foil cap)

Sterile tips for automatic pipettor

Universal bottles or suitable sterile germination vessel

Spatula

Forceps

Scalpel

Cork borer*

Small dissecting scissors.*

*Alternatives depending on tissue to be prepared.

Non-sterile:

Binocular dissecting microscope

Automatic pipettor.

Materials and reagents

Commercial bleach diluted to contain 6% available chlorine; this can also be made up from an aqueous laboratory preparation of sodium hypochlorite. Do not autoclave

Absolute alcohol

Surfactant (e.g. Tween 80)

Sterile distilled or deionized water

Germination medium (e.g. MS medium with 2% sucrose, 1.0 mg l^{-1} thiamine, 0.5 mg l^{-1} pyridoxine, 0.5 mg l^{-1} nicotinic acid, 0.5 g l^{-1} 2-(N-morpholino) ethanesulphonic acid (MES) pH 5.7 with 1 M KOH, 0.8% Bacto agar). Sterilized by autoclaving

Source of plant material (e.g. leaf or seed).

Protocol Part A – Germinating seedlings for callus production

1. Place the seeds in the tea strainer, make a small muslin bag to hold them or, if large, place them direct into a 250-ml conical flask.

2. Submerge them in absolute alcohol for a few seconds. Pipette off alcohol and discard.

3. Immerse the seeds in hypochlorite with a few drops of surfactant for 15–20 min (the duration of immersion can be varied – less for more sensitive seeds, longer if infection is a problem).

4. Rinse the seeds 3 × 5 min with 250 ml of sterile distilled water. If seeds float to surface, either pipette off liquid or try a low-speed centrifugation step (e.g. 1000 g for 5 min).

5. Working in aseptic conditions, place the seeds in a sterile Petri dish, then transfer them to germination medium in either a Universal bottle or other suitable sterile germination vessel. For *Arabidopsis*, pH appears to be very important; we always germinate it in a medium of less than pH 6. Seeds may be transferred by using an automatic pipette with the end of the tip removed.

6. The seedlings can then be germinated in a suitable incubator (temperature and lighting will depend on species; both species requiring a 16 h light : 8 h dark cycle and 22°C for *Arabidopsis* and 25°C for *Nicotiana*) and used to provide sterile material from which callus can be regenerated.

Protocol Part B – Explants from plant material for callus production

This method will work with a range of plant parts from plants grown in a growth chamber, glasshouse or in the field. Stems, leaves and other parts may all be regenerated. Root material presents more of a problem unless grown under aseptic conditions (see above) though cores bored from tubers or large roots using a sterile cork borer can produce successful callus.

1. Wash the plant material in water and dip it briefly in absolute ethanol.

2. Immerse it in hypochlorite with a few drops of surfactant added for 5–10 min (the duration of immersion can be varied – less for more sensitive material, longer if infection is a problem).

3. From now on, work in a laminar flow cabinet.

4. If necessary, use a sterile scalpel to cut away any surface material and repeat step 2; for instance, removing outer leaves or trimming away cut ends will all help to remove contaminants. Work on a sterile surface, like the inside of a Petri dish lid.

5. Rinse the tissue 3 × 5 min in sterile distilled water.

6. Now take the tissue and cut away the outer layers that have been in contact with the hypochlorite. Work on a sterile surface, using freshly sterilized instruments.

7. Cut the tissue into small sections up to 10 mm diameter and 2–3 mm thickness and place them onto callus medium (see below).

Protocol Part C – From embryos from imbibed seeds

This protocol is described for a gymnosperm, Norway spruce; however, it can be used for any seed large enough to be dissected.

1. Imbibe the seeds in water overnight. While mature seeds can be used, immature seeds may yield more viable cultures.

2. Surface sterilize the seeds, either in 5% sodium hypochlorite for 10 min, or 15% hydrogen peroxide for 15–20 min.

3. Rinse the seeds 3 × 250 ml of sterile distilled water.

4. Using aseptic technique and a binocular microscope carefully dissect the seeds and excise the embryo by squeezing the sides of the seed with fine forceps. The embryo will slip out of the endosperm and can be picked up with the fine forceps.

Embryos are ready for callus culture without further sterilization; between 10 and 20 are required per agar plate.

Notes

A very similar method can be used for maize callus; however, it is important to obtain embryos 12.5–17.5 mm long, 9–11 days after pollination. The entire ear is sterilized by immersion in sodium hypochlorite as above. The kernel is cut from the silk scar to the base and the embryo removed and placed onto callus induction medium (see below).

Protocol 6.2

allus from a dicot root: Arabidopsis thaliana and Nicotiana tabacum

Equipment

Sterile:

Petri dishes

Spatula

Scalpel.

Non-sterile:

Growth cabinet (22°C, continuous white light or 16 h light, 8 h dark)

Parafilm™ (Pechiney Plastic Packaging, Menasha, WI, USA) or 3M (3M Corporation) Micropore gas-permeable tape (this allows gas exchange and enhances growth).

Materials and reagents

Callus induction medium (sterile, in 6 cm Petri dishes)

Gamborg's B5 basal medium (Sigma G5768) with 2% glucose, 0.8% Agar, 0.5 g l^{-1} MES, 0.5 mg l^{-1} 2,4-D, 0.05 mg l^{-1} kinetin, pH 5.7.

(See Chapter 5 for instructions on preparing media)

Plant material – seedlings germinated in aseptic conditions; see method I above.

Protocol

(Work in a laminar flow hood):

1. Transfer seedlings to a sterile surface (e.g. the inside of a Petri dish lid).

2. Excise the roots and section them into roughly 1-mm lengths

3. Carefully transfer the cut pieces to the surface of the callus induction medium using sterile spatula.

4. Seal the Petri dishes around the edge with a piece of Parafilm™ or gas-permeable tape and place in an incubator at 22°C with continuous white light or 16 h light, 8 h dark.

5. Monitor over 2–3 weeks. Growth of callus will reach 3–5 mm over this period.

6. Callus may be subcultured by removing it and placing onto fresh medium periodically (every 4–6 weeks). It may be subdivided using a sterile scalpel to generate more individual calli.

Protocol 6.3

Callus from dicot shoot/leaf — Arabidopsis thaliana and Nicotiana tobacum

Equipment

Sterile:

Petri dishes

Spatula

Scalpel.

Non-sterile:

Parafilm™ or 3M Micropore gas-permeable tape

Growth cabinet (22°C or 25°C depending on species, continuous white light or 16 h light : 8 h dark cycle).

Materials and reagents

Arabidopsis callus induction medium as in *Protocol 6.2* (sterile, in 6-cm Petri dishes)

Nicotiana callus induction medium (sterile, in 6-cm Petri dishes)

MS medium with 3% sucrose, 0.8% agar, 1 mg l^{-1} 2,4-D, 1 mg l^{-1} kinetin, pH 5.2.

(See Chapter 5 for instructions on preparing media)

Plant material — surface sterilized leaves or seedlings germinated in aseptic conditions as *Protocol 6.1A* above.

Protocol

(Work in a laminar flow hood)

1. Excise 5–10-mm diameter segments of leaf material from expanding leaves.

2. Carefully transfer the cut pieces to the surface of the callus induction medium using a sterile spatula (it does not matter which way up the leaf pieces are placed).

3. Seal the Petri dishes around the edge with Parafilm™ or gas-permeable tape and place in an incubator at 25°C (for *N. tabacum*) and 22°C (for *Arabidopsis*) with continuous white light or a 16 h light : 8 h dark cycle.

4. Monitor over 2–3 weeks. Callus will begin to form at the edge of the cut leaf; it will be ready for reculturing on the same medium after 6 weeks and can be maintained by subculturing every 4 weeks. Callus may be subdivided using a sterile scalpel to generate more individual calli.

Protocol 6.4

Callus from a monocot (e.g. maize, rice)

While successful callus may be generated from roots and shoots, callus derived from immature embryos has the best potential for embryogenesis and the production of suspension cultures for protoplasts for transformations. The method for obtaining immature embryos is described above (*Protocol 6.1C*). The protocol is given with variations for rice and maize.

Equipment

Sterile:

Petri dishes

Spatula.

Non-sterile:

Parafilm™ or 3M Micropore gas-permeable tape

Growth cabinet (25°C, dark, or seal Petri dishes in aluminium foil).

Materials

Induction medium (sterile, in 6-cm Petri dishes): 4 g l^{-1} (Chu et al., 1975) (N6) salts (Sigma C-1416), 1 ml l^{-1} Eriksson's Vitamin mix (1000 × stock made from Sigma E-1511 powder), 0.5 mg l^{-1} thiamine-HCl, 0.1 g l^{-1} vitamin-free casamino acids. (See *Protocol 5.3/Table 5.3* for preparation.) Supplement with 2 mg l^{-1} 2,4-D, 20 g l^{-1} sucrose, 0.8% agar, pH 5.3 (maize) or 4.4 mg l^{-1} 2,4-D, 50 g l^{-1} sucrose and 1.0% agar, pH 5.8 (rice).

Plant material – immature maize embryos from kernels, see *Protocol 6.1C* above or use embryos from dehulled rehydrated rice seeds.

Protocol

(Work in a laminar flow hood):

1. Separate the embryo from the endosperm and place onto callus initiation medium, with the embryo axis in contact with the medium, scutellar side up. Ten embryos may be placed per plate.

2. Seal plate edges with Parafilm™ or 3M gas-permeable tape and incubate in the dark at 25°C.

3. Maize: observe over 2–3 weeks; when a friable, soft embryogenic callus has formed, transfer it to the plates containing the same medium as previously. The callus formed has numerous small embryoids protruding from its surface. Rice: observe growth and transfer to fresh media every 10–14 days.

4. If embryoids begin to develop further, increase the 2,4-D concentration or decrease the time between subculturing to maintain the culture as callus.

Protocol 6.5

Callus from gymnosperms (Norway spruce)

Equipment

Sterile:

Petri dishes

Spatula.

Non-sterile:

Parafilm™ or 3M Micropore gas-permeable tape

Growth cabinet (20–25°C, dark, or seal Petri dishes in aluminium foil).

Materials and reagents

Norway spruce callus induction medium (sterile, in 6-cm Petri dishes)

0.5 strength Litvay's medium with 1% sucrose, 2 mg l^{-1} 2,4-D, 1 mg l^{-1} benzyl-adenine, 500 mg l^{-1} casein hydrolysate, 250 mg l^{-1} glutamine, pH 5.6, 0.8% agar.

(See Chapter 5 for instructions on preparing media)

Plant material – embryos from imbibed seeds, see *Protocol 6.1C* above.

Protocol

(Work in a laminar flow hood):

1.   Place 10–15 embryos onto callus induction medium using a sterile spatula.

2.   Seal the Petri dishes around the edge with a piece of Parafilm™ or gas-permeable tape and place in an incubator at 20–25°C in the dark.

3.   Transfer to fresh medium after 2–4 weeks.

Monitor weekly; the culture can be maintained by subculturing every 4 weeks. Callus may be subdivided using a sterile scalpel to generate more individual calli that are placed onto new agar plates.

Protocol 6.6

Transforming maize callus by particle bombardment

Equipment

Sterile:

Petri dishes

Spatula

Scalpel.

Non-sterile:

Parafilm™ or 3M Micropore gas-permeable tape

Growth cabinet (22°C, continuous white light or 16 h light : 8 h dark cycle)

Particle gun (e.g. Biorad).

Materials and reagents

These media must all be sterile.

Chu N6 agarose medium (see *Protocol 6.4*, substituting agarose for agar; sterile in 6-cm Petri dishes)

Selection media: agarose plates as above supplemented with selection antibiotic or herbicide for the selectable marker.

Microprojectile/DNA preparation, including a selectable marker in the expression cassette (see Chapter 4).

Maize callus, less than 1 week old.

Protocol

(Work in a laminar flow hood):

1. Prepare a circle of embryogenic callus about 5 cm in diameter on Chu N6 agarose in a 25-cm Petri dish. Try to make the layer as thin as possible as the particles will only penetrate the upper cell layers adequately.

2. Place the Petri dish under the particle gun, align it with the callus and bombard it with microparticles according to the manufacturer's instructions. Optimization may be usefully carried out using a transient expression system.

3. Place the callus onto a selection medium and leave for 1–2 weeks (growth conditions as described for *Protocol 6.4* callus).

4. Transfer the callus to fresh plates with selective media. Grow for 1–2 weeks further. One initial plate should be spread thinly onto a number of new plates.

5. Some calli will grow better than the others in the presence of the selection medium. These calli are transgenic, and contain both the selectable marker gene and the gene of interest. Place them onto fresh selective plates and grow for a further 4 weeks. Confirm the presence of the transgene at this stage by molecular techniques.

6. At this stage, after 10–12 weeks, the calli are ready to be plated out for the regeneration of plants (Chapter 11) or for fragmenting, to form transgenic suspension cultures (Chapter 7).

Protocol 6.7

Preparation of a transgenic Arabidopsis thaliana callus using Agrobacterium

Equipment

Sterile:

Petri dishes

Spatula

Forceps

Scalpel.

Non-sterile:

Growth cabinet (22°C, continuous white light or 16 h light : 8 h dark)

Parafilm™ or 3M Micropore gas-permeable tape.

Materials and reagents

All media need to be sterile

1. Half strength MS liquid medium, 2% sucrose (to calculate, will need 25 ml per Petri dish. If using 10 plates prepare 250–300 ml).

2. Ten plates without antibiotics – half strength MS medium, 2% sucrose, 1% agar, 0.8 mg l^{-1} BAP, 0.1 mg l^{-1} IBA.

3. Ten plates as 2 above with antibiotics; 100 μg ml^{-1} carbenicillin, 20 μg ml^{-1} timentin (ticarcillin/clavulanic acid, Melford Laboratories) and selective antibiotics (e.g. 40 μg ml^{-1} hygromycin).

4. Ten plates of callus induction medium as *Protocol 6.2* above plus antibiotics as 3.

Protocol

(Work in a laminar flow hood):

1. Prepare callus leaf pieces as described in *Protocol 6.3* above.

2. Float the pieces on 25 ml of MS medium + 2% sucrose + 400 μl 24 h *Agrobacterium* culture in a sterile Petri dish.

3. Gently agitate for 20 min at 28°C and 50 rpm in an orbital incubator in the dark.

4. Take squares out of liquid medium and place onto antibiotic-free plates leaving a 1-mm gap between leaf pieces.

5. Wrap plates in foil.

6. Incubate for 3 days at 22°C in the dark.

7. Move the leaf squares to antibiotic plates and seal edges with Parafilm™ or 3M Micropore gas-permeable tape.

8. Transfer leaf squares to fresh antibiotic plates at weekly intervals. If shoot formation is desired, continue to use the plates described under Materials and Reagents item 3 above. If callus production is desired, use the plates described in Materials and Reagents, item 4 above.

9. Subculture the calli at regular intervals to fresh plates. Generation of plants from callus is described in Chapter 11.

Chapter 7

Cell suspension cultures

1. Introduction

Cell suspension cultures are rapidly dividing suspensions of cells grown in liquid medium. In general, suspension cultures grow more rapidly than callus cultured on agar and are more amenable to experimental manipulation. The ideal suspension culture consisting entirely of single cells, which will allow the application of standard microbial culture techniques, is rarely ever achieved. Most suspension cultures are comprised of cell aggregates as well as dispersed single cells. However, by selecting and subculturing for several generations, a fine cell suspension culture consisting of a dispersion of single cells and small cell aggregates can be established and maintained.

2. Initiation of cell suspension cultures

The time required to initiate and establish a cell suspension culture depends on the species of plant and the growth medium. However, in general, dicot species are easier to establish in cell suspension culture than monocot species.

Cell suspension cultures are usually initiated by agitating a fragment of *in vitro* grown callus in a volume of liquid medium on an orbital shaker (*Protocol 7.1*). The essence of the procedure is to disperse callus into single cells and small masses of cells. Three main procedures are used to achieve dispersion:

- initiation from friable callus;
- initiation from non-friable callus;
- initiation from callus treated with cell wall degrading enzymes.

In all cases, the callus selected for initiation of a suspension culture should be healthy and vigorously growing. The appearance of the culture is an important diagnostic feature; white or cream-coloured callus usually indicates a healthy culture whereas dark brown colouring suggests moribund or dead cells. To assess the condition of the culture at the cellular level, a little material may be taken aseptically and examined by light microscopy. The number of healthy cells can be quickly determined by staining with Evans Blue dye (see Chapter 7, *Protocol 7.3*), or the fraction of dividing cells can be estimated from the mitotic index by fluorescent microscopy with Hoechst stain. The most reliable measure of the viability and vigour of the callus culture is the growth rate.

2.1 Induction from friable callus

The most commonly used starting material for the initiation of suspension cultures is friable callus because it is easily fragmented during agitation in liquid culture. As friability is an important factor for the successful initiation of a suspension culture, a number of procedures can be used to obtain suitably friable callus. For example, the callus can be passaged on a 7-day cycle for 2–3 weeks immediately prior to its transfer to liquid culture and/or the ratio of auxin to cytokinin in the callus medium can be altered by raising the auxin concentration.

At initiation, it is also important to set up an appropriate ratio of callus tissue to liquid medium. Approximately 2–3 g of friable callus per 100 ml should establish a healthy suspension. Low levels of callus tissue will not replicate successfully. A denser seeding of callus can be monitored and passaged on at an early stage and is preferable to very low quantities of material in a large volume of liquid. As the cells break off from the callus and begin to form a suspension it will be necessary to subculture the cells to fresh medium; this should be kept to a ratio of 1:1. The procedure of repeated subculture should be continued until the culture reaches the desired density and is actively growing.

Actively dividing fine cell suspensions can be selected for at the early stages of culture initiation by filtering to remove the larger aggregates. By repeated filtering and subculturing, the culture can be reduced to a suspension of small aggregates and free floating single cells.

2.2   Initiation from non-friable callus

Where the callus available for the initiation of the cell suspension culture is non-friable, fragments of callus can be transferred to liquid medium and grown with agitation on a shaker. The callus is subcultured regularly until it reaches a suitable degree of friability. The callus is then used to initiate and establish cell suspension cultures as described above for friable callus.

2.3   Initiation from callus treated with enzymes

In the early stages of initiation, callus is transferred and subcultured in a liquid medium containing low concentrations of cell wall degrading enzymes until a cell suspension is formed. Pectinase, which breaks down the middle lamella of the plant cell wall and separates plant cells, is frequently used and cellulase is sometimes added.

3.   Maintenance of cell suspension cultures

Cell suspension cultures can be grown asynchronously either in batch culture or in continuous culture. A variety of culture vessels and shakers are available commercially that can be used for the batch culture of plant cell suspensions (*Figure 7.1*). The most suitable vessel for batch is one that allows a large surface area to maximize gas exchange. For example, 20 ml of culture medium in a 100-ml conical flask or 50–100 ml in a 250-ml flask should allow sufficient gas exchange when the culture is being shaken at 100–120 rpm and at 25°C (*Figure 7.2*). Aluminium foil caps, sterilized and dried, should be used to seal the flask. Conical flasks with internal baffles are sometimes used to fragment any aggregates that might form. Established cultures are subcultured every 1–3 weeks depending on the growth of the culture. Transfers are made under aseptic conditions by using a wide-bore pipette to prevent blocking of the outlet. Suspension cultures should be

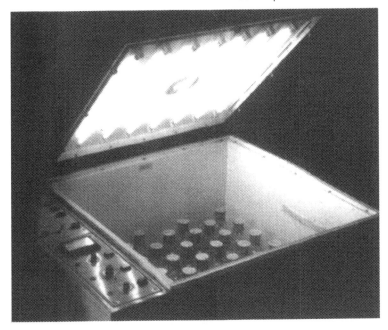

Figure 7.1

A typical orbital incubator shaker for suspension cultures. Illumination is provided by banks of horticultural quality fluorescent tubes in the lid of the incubator (shown open) and the flasks are shaken with an orbital motion on a platform in the base. Temperature is controlled on a day/night cycle with the lighting.

subcultured in early stationary phase (*Figure 7.3*) and at regular intervals, the embryogenic capacity of the culture should be assessed.

Several different types of fermentor or bioreactor can be used for the continuous culture of plant cell suspensions. The most common type of system for use in the laboratory is a stirred-jar fermentor, as used for microbial cultivation, with a culture volume of around 7.5 l. Cell cultures grown commercially for the production of technical compounds and polymers are cultivated in sophisticated bioreactors with capacities of 75 000 l.

4.   Growth characteristics of cell suspension cultures

4.1   Types of cell

Plant cell suspension cultures normally consist of cells with diverse morphology and state of aggregation. Cells in aggregates exist in different microenvironments from single cells and respond differently to changes in the culture environment. In many suspension cultures, at least two morphological types of cells can be distinguished. Cell aggregates are normally made up of small isodiametric cells. Single cells may be large and elongated depending on the type of auxin used (*Figures 7.4* and *7.5*). The proportion of the cell types changes during the passage of the culture and depends on the nature of the auxin present in the culture medium.

Figure 7.2

A flask of suspension culture cells of carrot. A larger flask (250 ml) than the volume of suspension (100 ml) is required to ensure sufficient aeration.

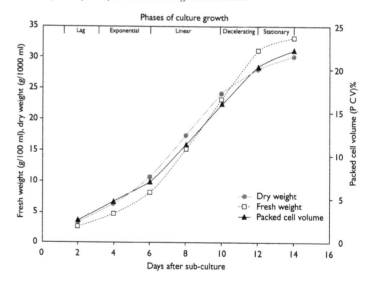

Figure 7.3

Growth curves of carrot cells in suspension culture. In this case, the entire culture cycle lasts about 14 days, with the cells entering the stationary phase at the end of this period. To maintain the culture, cells should be subcultured to fresh medium when in the early stationary growth phase.

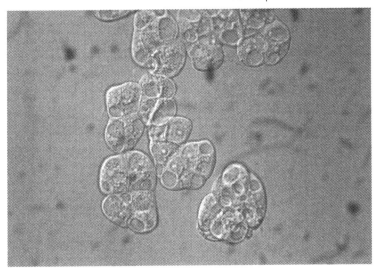

Figure 7.4

Isodiametric cells of carrot obtained in suspension culture in the presence of 2,4-D. Note the cell clumps formed as cells divide but do not separate.

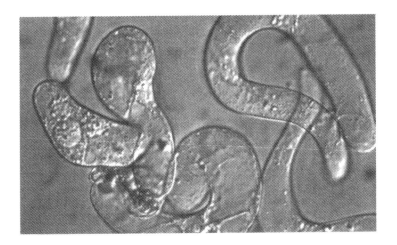

Figure 7.5

Elongate cells of carrot obtained in suspension culture in the presence of NAA.

4.2 Monitoring the growth of the culture

Growth of cell suspension cultures can be followed by measuring one or more of the following parameters at intervals during the growth cycle:

- cell number;
- packed cell volume (PCV);
- fresh weight and dry weight;
- cell viability;
- medium conductivity.

It is good practise to choose two methods and to use these to measure the culture until it appears to have entered stationary phase.

Cell number

The number of cells per unit volume can be determined using a haemo-cytometer. The suspension is diluted so that at least three fields can be scored and the total cells recorded should be greater than 1000. It should be noted that cell aggregates and the large size of single cells in some suspensions can cause incomplete loading of the counting chamber of the haemocytometer and consequently lead to unreliable cell counts. These difficulties can be overcome by dispersing the cell aggregates with 10% HCl and 10% chromic acid as described by Reinert and Yeoman (1982) and by using an improved Neubauer type haemocytometer (depth of counting chamber, 0.1 mm) to accommodate the larger cells. Great care should be taken when using corrosive acids and the acid treatments should be carried out in a fume hood.

Packed cell volume (PCV)

Aliquots (10–15 ml) of evenly mixed cell suspensions are transferred to a graduated tube and spun at low speed in a centrifuge fitted with a swing-out rotor. The PCV is calculated as the percentage total volume occupied by the cell pellet (see *Protocol 7.2*). The pellet can be macerated and the cell count determined as described above. PCV is accurate only for fine suspension cultures.

Fresh weight and dry weight

Fresh weight is determined by weighing freshly harvested cells and dry weight from cells dried at 60°C for 48 h (see *Figure 7.3* and *Protocol 7.2B*).

Cell viability

Rough estimates of the percentage of viable cells in a culture can be made from cell counts by using viability stains like Evans blue and fluorescein diacetate (FDA; see *Protocol 7.3A* and *7.3B*)

Conductivity and pH

The conductivity of the growth medium is sometimes used as a convenient method of monitoring the growth of a suspension culture. Conductivity measurements can be made rapidly with a conductivity meter, but each new suspension culture must be assessed for the accuracy of this technique by comparison with other growth data. In general, the conductivity of the media falls over the passage and probably reflects the uptake and utilization of electrolytes from the medium.

Changes in pH also reflect the various phases of the culture cycle. Cultures become very acidic early on in logarithmic phase (pH 3.5–4.5) and return to approximately pH 5 in stationary phase. The pronounced drop of pH early in the growth cycle may be due to proton production from meta-bolic activity and the rapid removal of media constituents like phosphate that provide the buffering capacity of the medium.

4.3 Contamination

It is essential to check cultures before and soon after subculturing for any signs of microbial contamination. An opaque, dense white or pink color-ation that appears very rapidly (i.e. within 24 h) is usually indicative of a yeast or bacterial contamination. The identity of the contaminant can be confirmed by examining a sample of the culture under the microscope. Fungal contamination usually appears as balls of mycelium that develops 48 h after transfer. All cultures suspected of contamination should be disposed of immediately by autoclaving.

5. Uses of cell suspension cultures

Plant cell suspension cultures have found wide application both for research and for commercial exploitation.

For research, cell suspension cultures are easy to maintain and allow great flexibility of experimental approach. Cell suspensions are ideal to study various factors and compounds that affect growth and differentiation. One of the key advantages is the feature that most of the cells are in direct contact with the medium and therefore the effects of concentration gradients are avoided. Cell suspension cultures provide a useful system to study cell division in the absence of developmental processes and their growth can be made synchronous for studies on the cell cycle. Another important feature of suspension cultures is that they can be used for the rapid preparation of protoplasts in high yield. Protoplasts themselves are a unique system for basic and applied research.

The formation of somatic embryos in suspension cultures is ideal for the large-scale production of commercial plants. The greatest potential commercial application of suspension cultures is their use as a production system for plant-derived chemicals and recombinant pharmaceutical proteins (see Chapters 14 and 15).

Reference

Reinert, J. and Yeoman, M.M. (1982) *Plant Cell and Tissue Culture: A Laboratory Manual*. Springer-Verlag, Berlin.

Protocol 7.1

A generic protocol for initiating a suspension culture

This is a protocol suitable for different types of callus.

Equipment

Incubator set at correct temperature for callus (see Chapter 6)

Orbital incubator, shaking at 130 rpm, set at correct temperature for species (see Chapter 6), fitted with fluorescent lights and timer

Laminar flow cabinet

Bunsen burner

Autoclave

Scalpel and forceps, autoclaved in aluminium foil

Glass Petri dish (to hold tools while cooling)

Sterile glass or plastic Petri dishes.

Materials and reagents

Petri dishes of solid medium containing calli to be transferred to suspension cultures

Flasks of medium appropriate to the species of calli (see *Protocols 6.1–6.5*). Volumes of 20 ml in a 100-ml flask or 50–100 ml in a 200-ml flasks are recommended

70% w/v alcohol.

Protocol

1. Thoroughly sterilize the laminar flow cabinet with 70% alcohol. Leave running for 20 min before use.

2. Spray foil-wrapped tools with 70% alcohol and leave to dry in the cabinet.

3. When bringing plates of callus into the cabinet, check carefully that the marker pen used to label the plates does not wash off in the presence of alcohol. If it does, wipe unmarked areas of plates with paper soaked with alcohol.

4. Spray flasks of medium with 70% alcohol and allow to dry.

5. Unwrap sterile tools and Petri dishes of callus.

6. Using forceps and scalpel move callus to sterile Petri dish and break apart into small pieces. Ideally, the tissue should be approximately 2–3 mm in diameter. Close Petri dish.

7. Resterilize the tools in a bead sterilizer. Allow to cool before use.

8. Remove foil cap from fresh flask of medium.

9. Using forceps, pick up callus and transfer into flask. Avoid touching sides of neck with callus.

10. Seal neck of flask with a fresh foil cap.

11. Move flask to orbital shaking incubator. Monitor growth of material by eye. Discard contaminated flasks. Within a week or two it should be possible to observe an increase in the number of free cells in the medium which have broken away from the callus.

12. After 7–10 days, the culture should be passaged, whether or not a large increase in mass has occurred. If the volume of tissue has not increased greatly, use a smaller volume of medium for the second passage.

13. Allow the material to settle at the bottom of the flask. Pipette off the liquid medium and replace with fresh.

14. Repeat this method of subculture until a reasonable mass of material has accumulated. Then pipette a volume of cells (for example 10 ml into a 100 ml) from the old flask into a fresh flask.

15. Should the material not be increasing in mass at an appreciable rate, repeat the procedure from the beginning using a larger volume of dispersed callus to the same volume of liquid medium.

Protocol 7.2

Monitoring the growth of suspension cultures

Part A – Packed cell volume (PCV)

Equipment

Sterile 10-ml pipette

Automatic pipette (pipette filler)

Sterile, graduated plastic centrifuge tube

Bench centrifuge (swing-out rotor preferable)

Graph paper.

Materials and reagents

Flasks with equal volumes of suspension culture cells, subcultured from a single source flask. There should be one flask for each day of the planned time course.

Protocol

1. Pipette a fixed volume of suspension culture into a 15-ml graduated plastic centrifuge tube and either allow cells to settle out by gravity or centrifuge at 1000 g for 5 min. Ensure the flask is thoroughly shaken immediately before sampling as suspension culture cells settle very rapidly.

2. Record the volume of the pellet.

3. Repeat this every day at the same time and plot against time. The numbers should show a rapid increase during the logarithmic phase of growth and then reach a plateau (stationary phase). This should be repeated for the new culture when its growth cycle is stable. With this information, establish a day when the culture is in early stationary phase and choose this as the routine day for subculturing.

Part B – Measurement of fresh weight and dry weight

Equipment

Glass filter funnel

Whatman no. 1 filter paper, 9 cm in diameter

Buchner flask (with side arm for vacuum attachment)

Fine balance

Oven set to 60°C.

Materials and reagents

Flasks of cells. There should be a separate flask of cells for each day of the planned time course.

Protocol

1. To determine wet weight, filter the cell suspension through a pre-weighed Whatman no. 1 filter paper, allow it to drain under vacuum and reweigh the filter.

2. To determine a dry weight, use a pre-weighed filter paper, filter a known volume of cells through it and dry in a 60°C oven for 48 h. Because dried cells weigh very little this method is likely to be inaccurate unless large volumes are filtered.

3. Repeat with replicates to obtain an accurate growth curve. This should be repeated at the same time every day until no further growth is detected and the culture has reached stationary phase.

4. Plot the weights, wet or dry against time to produce the growth curves.

Protocol 7.3

Measurement of cell viability

Part A – Exclusion of Evans blue dye

Equipment

Pipette

Light microscope

Small plastic tube, 1.5 ml in capacity

Slides and coverslips.

Materials and reagents

Flask of suspension-cultured cells to be assayed

Stock solution of Evans blue stain, 0.1% w/v in water.

Protocol

1. Mix 1 drop of stock Evans blue dye solution with 0.5 ml of cell suspension in a small plastic tube.

2. Allow to incubate for 2 min at room temperature.

3. Transfer 1 drop to a slide and apply a coverslip, taking care to avoid bubbles.

4. View under the light microscope. Ensure that there is a minimum of 100 cells in the field of view. Living cells exclude Evan's blue dye, while dead cells are stained blue.

5. Count the numbers of stained and unstained cells. Calculate the percentage of living cells as a measure of viability.

Part B – Fluorescent staining with fluorescein diacetate (FDA)

Equipment

Fluorescence microscope

Slides and coverslips

Automatic pipettes, 0–10-μl and 200–1000-μl capacity

Eppendorf tube, 1.5-ml capacity.

Materials and reagents

FDA (Sigma Chemicals, Poole, Dorset) stock solution prepared as 0.5 mg ml^{-1} dissolved in acetone.

Cell or protoplast suspension.

Protocol

1. Mix 5 μl of stock FDA solution with 0.5 ml of protoplast suspension in an Eppendorf tube.

2. Incubate for 2 min at room temperature.

3. Place a drop on a slide, cover with a coverslip and view under a fluorescence microscope. Ensure that there are at least 100 cells visible and that they are separate from each other.

4. Count the fluorescent and non-fluorescent cells in the field of view. The fluorescent cells are viable. The percentage of viable cells should exceed 80% in a healthy suspension culture.

Part C – Mitotic index with Hoechst stain

Equipment

Fluorescence microscope

Slides and coverslips

Automatic pipettes 0–10 μl and 200–1000 μl capacity

Materials and reagents

Hoechst stain 1 mg ml^{-1} in water (wear gloves when handling)

Protoplast or cell suspension.

Protocol

1. Add 1 drop of Hoechst stain to a drop of cells on a microscope slide.

2. Incubate for 5–15 mins.

3. Apply coverslip and gently absorb excess stain with tissue.

4. Observe by fluorescence microscope with U.V. filter set. Dividing nuclei will fluoresce strongly blue. Mitotic index can be estimated from the proportion of cells undergoing division.

Protocol 7.4

Genetic transformation of suspension cultured cells

Equipment

Small Petri dish

Automatic pipette and sterile tips

Incubator, without lights, set at 25°C

Bench centrifuge and tubes

Sterile forceps and scalpel.

Materials and reagents

Flask of suspension cultured cells, 3 days old

Agrobacterium tumefaciens overnight culture (OD at 600 nm reading 2.0)

Fresh, sterile cell culture medium

Petri dishes of cell culture medium plus 0.8% agar and supplemented with a selecting agent such as an antibiotic (e.g. kanamycin or hygeromycin, see Chapter 4) and an antibiotic to kill the *Agrobacteria* (e.g. carbenicillin and timentin, see Chapter 6).

Protocol

1. Take 1 ml of a 3-day-old cell suspension of cells and transfer to a small, sterile Petri dish.

2. Add 50 μl of an overnight culture of *Agrobacterium tumefaciens*.

3. Mix the culture and incubate in the dark for 2 days at 25°C.

4. Aseptically transfer by pipette to a centrifuge tube and spin at 50 *g* for 5 min.

5. Wash pellet three times with sterile culture medium to remove excess, unattached bacteria.

6. Take plates of solid culture medium supplemented with appropriate selecting antibiotic and antibiotic mix to remove *Agrobacteria*.

7. Aseptically pipette 1 ml of the culture on to each plate and spread cells across the entire surface by shaking.

8. Incubate for 1 month at 25°C in the dark to allow microcalli to develop on the plates.

9. After this time, screen the plates for transformed microcalli (for example, if using GFP expression, screen with an ultraviolet lamp).

10. Transfer the selected microcalli, using sterile forceps, on to fresh plates.

Chapter 8

Protoplast culture

1. Introduction

The plant cell wall is a multi-layered structure composed of polysaccharides and proteins. The polysaccharides are cellulose, a polymer of glucose; pectic compounds, which are polymers of galacturonic acid molecules; and hemi-cellulose, a polymer consisting of a variety of sugars including xylose, arabinose and mannose. Cell walls consist of three types of layers.

■ The middle lamella – the first layer formed during cell division, forming the outer surface of the wall and is shared by adjacent cells. It is composed of pectic compounds and proteins.

■ The primary wall – is formed after the middle lamella consisting of a framework of cellulose microfibrils embedded in a gel-like matrix of pectic compounds, hemicellulose and glycoproteins.

■ The secondary wall – formed after cell enlargement, a rigid structure composed of cellulose, hemicellulose and lignin.

Cell walls fulfil several important roles, but two key functions are: (i) to provide tensile strength and limited plasticity so that the cell can develop high turgor pressures without rupturing. Turgor pressure provides support for non-woody plants. (ii) To provide a tough physical barrier that protects the interior of the plant cell from invading micro-organisms. The pore sizes of the wall are small enough to exclude even the passage of viruses. Microbes that are saprophytes and pathogens of plants have a range of hydrolytic enzymes that they use to degrade the cell wall and gain entry to the cell's interior.

The properties of the plant cell wall that make it an efficient protective barrier *ipso facto* restrict the use of plant cells for a variety of cell and tissue culture techniques, including the delivery of large molecules into the cell, somatic hybridization and genetic manipulation. These techniques can therefore only be applied to plant cells after removal of the cell wall.

Careful removal of the plant cell wall results in a viable, spherical cell called a protoplast (*Figure 8.1*). Protoplasts consist of the original cell's contents bounded by the plasma membrane. With appropriate culture conditions protoplasts will resynthesize the cell wall and undergo cell division to form callus from which new plants can be regenerated.

2. Isolation of protoplasts

Protoplasts can be isolated from a variety of whole-plant tissues and from plant tissue cultures like callus and suspension cultures. Removal of the cell

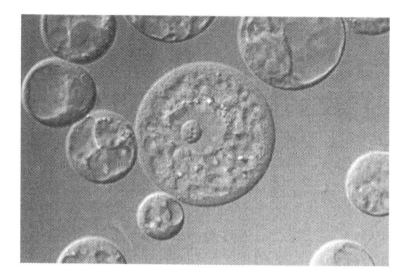

Figure 8.1

Protoplasts isolated from carrot suspension cultures by enzymatic digestion of the cell wall.

wall for protoplast isolation is achieved either by mechanical means or by enzymatic digestion.

2.1   Mechanical isolation

The essence of the mechanical technique is to subject plasmolysed tissue to a number of sharp cuts followed by de-plasmolysis to release the protoplast from the cut ends of cells. This technique, which was one of the earliest methods attempted, is difficult to apply and the yield of protoplasts is meagre. For practical reasons, this method is rarely used except for special situations where wall-degrading enzymes have an unavoidable deleterious effect on the protoplasts and only small quantities of protoplasts are required.

2.2   Isolation by enzymatic digestion of the cell wall

Isolation of protoplasts by the enzymatic digestion of the cell wall was first introduced by Cocking in 1960 and is now the universally adopted method. A combination of cellulases, hemicellulases and pectinases are used to break down the cell wall. The enzymes are obtained from culture filtrates of micro-organisms that degrade cell wall material. A commonly used cellulase (Onozuka R10) comes from the filamentous fungus *Trichoderma reesei* and contains an assortment cellulases and hemicellulases. The most frequently used pectinase is Macerozyme and alternatives are Pectolyase and Pectinase (Sigma, St Louis, Missouri, USA).

The most important factor in the preparation of intact and viable proto-plasts is to maintain isotonic conditions during isolation in order to prevent osmotic lysis of the protoplasts in the absence of the supporting cell wall. Usually, the tonicity of the preparation medium is maintained by adjusting the concentrations of inert osmostabilizers such as mannitol and sorbitol. However, in some procedures, metabolizable sugars such as sucrose are also added.

The general procedure involves incubating the plant tissue in a salt solution containing the hydrolytic enzymes and a suitable osmostabilizer (see *Protocol 8.1*). The digestion medium contains calcium ions to aid membrane stability and a zwitterionic buffer to control pH, especially pH changes that may occur as a result of cell lysis. Generally, the concentration of the mannitol or sorbitol is such that incipient plasmolysis (shrinking of the protoplasts from the cell wall) occurs. It is not uncommon to expose the source tissue to a preliminary plasmolysis incubation step before addition of the wall-degrading enzymes. However, this step is not included in some recent procedures, particularly those requiring long enzymatic digestions.

When the source tissue is stem or root, the material is cut into sections. Leaves are cut into strips, or for leaf mesophyll protoplasts, the epidermis is removed with the aid of tweezers and the stripped leaves are then exposed to the incubation medium. Penetration of the digestion medium into the tissues is sometimes helped by infiltration under vacuum. If the source is callus, the material is broken into small pieces. Suspension cultured cells are harvested, washed and resuspended in the appropriate incubation medium. The length of time the tissue is exposed to the hydrolytic enzymes depends on the tissue and on the concentration of enzymes used. With lower concentration of enzymes, incubations proceed for 12–14 h, whereas with higher concentrations of enzymes the total incubation time is usually 2–3 h.

2.3 Purification of isolated protoplasts

At the end of the digestion period, the isolated protoplasts have to be separated from the enzymes and cellular debris and transferred to a suitable medium. This purification step is accomplished by gentle centrifugation, filtration through a nylon mesh (approximately 60–70 μm) followed by centrifugation or by the use of a density gradient centrifugation step.

2.4 Protoplast viability and density

The viability of the isolated protoplasts can be determined by the same viability tests as used for cells in suspension culture. Exclusion of Evans blue dye detected by microscopy or accumulation of FDA visualized by fluorescent microscopy are standard viability tests (see Chapter 7).

The density of protoplasts in the suspension (number/unit volume) is usually determined by counting with a modified haemocytometer such as an improved Neubauer or Fuchs–Rosenthal with a field depth of 0.2 mm.

3. Protoplast culture

Isolated protoplasts can be cultured in an appropriate medium to reform cell walls and generate callus. In determining the optimal culture conditions, it is important: (i) to determine the optimal density for culture. If the density is too low, the protoplasts will lose soluble cell components because of the relatively high volume of surrounding medium and fail to divide. (ii) To establish the optimal auxin to cytokinin ratio. This can be determined by the growth square arrangement as described in Chapter 6. (iii) To maintain the osmoprotectant in the medium until the cell wall has reformed. The concentration can be reduced incrementally with successive subcultures as the cells grow and divide.

Protoplasts can be cultured in the following ways.

- Hanging-drop cultures. This involves culturing protoplasts in droplets (100 µl) suspended on the lid of a Petri dish, with sterile water in the base of the dish to provide humidity. Because of the small volumes required this arrangement is a convenient way of establishing optimal conditions of growth.
- In the wells of microtitre plates.
- On a semi-solid medium using agarose plates. This is an effective method as it provides a supporting matrix for the protoplasts. Standard agar is toxic to protoplasts and low-temperature gelling agarose is used instead. The protoplasts are either layered on the surface of the agarose or embedded in the agarose by suspending in a small volume of the gel at 40°C and added to a small Petri dish before it sets. Alternatively, the protoplasts–gel suspension is layered on a Millipore filter or a nylon membrane to set. The protoplasts embedded in the gel on the membrane are cultured over a nurse culture. The nurse culture normally consists of actively dividing non-embryogenic cultured cells embedded in agarose gel. The nurse culture is thought to provide the protoplasts with growth factors that stimulate wall synthesis and cell division.

Protoplasts from some species begin cell wall regeneration within a few hours of isolation whereas others may take several days to complete the new wall. The formation of the wall can be followed by using the compound Calcofluor White (see *Protocol 8.9*). This dye fluoresces when it binds to wall material and can be detected by fluorescent microscopy. Once the protoplasts have formed a new wall, they undergo cell division and callus is generated.

4. Uses of protoplasts

Protoplasts are a single cell system that is useful for basic research as well as applied plant science. Protoplasts can be used in plant cell metabolic studies including photosynthesis. They are ideal for studies on cell wall synthesis and deposition and are the starting point for the isolation of organelles like vacuoles and nuclei. In addition, protoplasts are the most amenable higher plant system for the application of flow cytometry. Flow cytometry is a powerful technique that uses specific fluorescent labelling of cellular components and has found wide application in studies on mammalian cells. By using endogenous fluorescent molecules or added fluorescent tags, flow cytometry can be used for cellular analysis and cell sorting of plant protoplasts (Coba de la Pena and Brown, 2001).

Protoplasts can be transformed by *Agrobacterium*, and without the cell wall barrier they are a convenient cell system for genetic transformation by direct methods of DNA transfer. Direct transfer of plasmid DNA into protoplasts is achieved by:

- electroporation – transient pores are induced in the plasma membrane of protoplasts by short, high-voltage pulses. Some of the pores are large enough to allow the uptake of DNA and will reseal spontaneously trapping the DNA in the protoplast (see *Protocol 8.12*);

■ chemical procedures – based on the addition of calcium or PEG. The use of PEG methods also provides a useful way of directing DNA uptake into chloroplasts (*Protocol 8.10*).

Protoplast transformation by direct DNA transfer is a very effective strategy to obtain transient gene expression. The technique of transient expression is increasingly being used in basic research to study a range of plant biological problems, because it does not suffer the disadvantages of delay and environmental risks that are associated with stable transformation systems. However, in transformation work aimed at improving cultivars, techniques involving protoplasts are now avoided because they are labour intensive and the recovery of plants often requires long culture periods with the risk of generating somaclonal variants (Chapter 12).

4.1   Protoplast fusion

In plant breeding, the technique of somatic hybridization by protoplast fusion is a particularly useful way of introducing genetic variation from distant relatives into crop plants in circumstances where sexual hybridization is prevented by reproductive barriers. When two or more protoplasts fuse to form a new cell, the parent nuclei may remain separate or fuse to form a somatic hybrid (*Figure 8.2*). If one of the nuclei is lost after fusion, the cytoplasms of the two parent protoplasts will still coalesce and form what is known as a cytoplasmic hybrid or cybrid. Cytoplasmic hybrids contain the nucleus of one parent protoplast but a mixture of the cytoplasmic organelles of both parents. The production of cytoplasmic hybrids is an important technique in plant breeding programmes where CMS is desirable, because the CMS trait is inherited through mitochondria.

Isolated protoplasts can fuse spontaneously, although this occurs infrequently because of the mutual charge repulsion that results from the net negative surface charges on the protoplasts' plasma membranes. Nevertheless, fusion can be induced either by the application of compounds

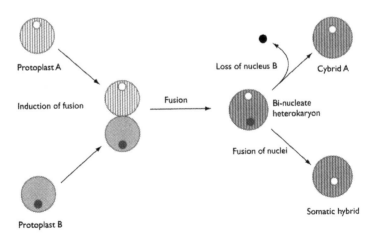

Figure 8.2

Scheme showing the formation of somatic hybrids by protoplast fusion.

known as fusiogenic agents or by placing the protoplasts in an electric field (electrofusion; *Protocol 8.11*).

Fusiogenic agents tend to either neutralize or shield the surface charge so that adjacent protoplasts come in to close proximity, adhere and coalesce. Fusion is induced by high concentrations of calcium ions at high pH (pH 10) and high temperature (37°C). However, the most widely used fusiogenic agent is PEG.

Electrofusion takes place in two stages. First, the protoplast suspension is placed in a fusion cell between two metal electrodes and subjected to a non-uniform alternating electric field. The protoplast membranes become differentially charged and positive regions are attracted to negative regions on adjacent protoplasts resulting in the protoplasts aligning in chains. In the second step, a high-voltage DC pulse is applied to the aligned protoplasts that cause breakdown of adjacent membranes and fusion.

Flow cytometry can be used to identify the fusion products by cell sorting and by estimation of nuclear DNA.

Further reading

Carlson, P.S., Smith, H.H. and Dearing, R.D. (1972) Parasexual interspecific plant hybridisation. *PNAS USA* **69**: 2292–2294.

Collins, H.A. and Edwards, S. *Plant Cell Culture*, pp 77–80. BIOS Scientific Publishers Ltd., Oxford.

Freeling, M. and Walbot, V. (Eds) (1993) *The Maize Handbook*. Springer-Verlag, Berlin.

Mathur, J. and Koncz, C. (1998) In: *Methods in Molecular Biology*, Vol. 82, Arabidopsis *protocols* (eds J.M. Martinez-Zapater and J. Salinas). Humana Press, Totowa, New Jersey, pp. 267–276.

Takebe, I., Labib, G. and Melchers, G. (1971) Regeneration of whole plants from isolated mesophyll protoplasts of tobacco. *Naturwissen* **58**: 318–320.

References

Coba de la Pena, T. and Brown, S. (2001) Flow cytometry. In: *Plant Cell Biology*, pp. 85–106 (eds. C. Hawes and B. Satiat-Jeunemaitre). Oxford University Press, Oxford.

Cocking, E.C. (1960) A method for the isolation of plant protoplasts and vacuoles from tomato root tip. *Nature* **23**: 29–50.

Protocol 8.1

Preparation of protoplasts from suspension cultures (e.g. maize) using a long incubation

Equipment

Sterile 250-ml conical flask

Bench centrifuge and centrifuge tubes

Pasteur pipette or automatic pipettor

Light microscope

Orbital shaker

Aluminium foil

Epifluorescence microscope with UV filter set

Sterile 50-ml Falcon tubes.

Materials and reagents

Solution A: 100 ml of culture medium supplemented with 0.6 M (16.4 g) sorbitol, 0.43 g MS basal salts, 3 g sucrose, 2 mg l^{-1} 2,4-D, pH 5.2

Solution B: 25 ml of Solution A supplemented with enzyme mixture: 0.25 g Cellulase Onozuka RS, 0.25 g Macerozyme R10 (Onozuka) and 0.05 g pectolyase

Flask of suspension culture cells judged to be in the logarithmic phase of growth (usually 2–3 days after subculture, see Chapter 7).

Protocol

1. Spin down 50 ml of suspension-cultured cells at 1000 g for 5 min.

2. Resuspend in 25 ml of Solution B and transfer to a sterile vessel that will allow maximum surface area (e.g. 30 ml of cell suspension in the bottom of a 250-ml conical flask).

3. Incubate at 50 rpm on an orbital shaker overnight at 25°C. The flask should be covered in aluminium foil to keep the enzyme solution dark.

4. Transfer to a sterile 50-ml Falcon tube and centrifuge at 600 g for 5 min.

5. Remove the supernatant with a pipette or a fine needle and discard.

6. Resuspend the protoplast pellet in 25 ml of Solution A using a wide-mouthed instrument such as a cut automatic pipette tip to reduce shear. Alternatively, add wash solution and partially invert tube very slowly.

7. Repeat centrifugation step once more. This should ensure that the enzymes are removed from the protoplasts. Finally, resuspend pellet in 5–10 ml of Solution A.

8. Make an initial check by light microscope for physical appearance (intact protoplasts are spherical) followed by FDA uptake using a fluorescence microscope (see *Protocol 7.3*). Cells that appear fluorescent green are viable.

Protocol 8.2

Isolation of protoplasts from carrot suspension cultures using a rapid incubation

Equipment

As *Protocol 8.1* with the addition of sterile glass wool.

Materials and reagents

Flask of cells judged to be in logarithmic phase of growth (see Chapter 7)

Solution A: 100 ml: 0.43 g MS medium, 2.5 g sucrose, 5 ml of coconut water, 100 µg l^{-1} of 2,4-D, 100 µg l^{-1} zeatin, 0.4 M (10.93 g) sorbitol, pH 5.0

Solution B, 25 ml: As solution A, but supplemented with 1 g Onozuka R10 cellulase, 0.5 g pectinase (Sigma).

Protocol

1. Take 50 ml of cells and centrifuge at 750 *g* for 10 min at 25°C.

2. Wash twice in 25 ml of Solution A and transfer to sterile 250-ml conical flask. Cover flask in aluminium foil to keep contents dark.

3. Shake on an orbital shaker at 50 rpm for 10 min, 25°C to ensure plasmolysis.

4. Repeat centrifuge step.

5. Resuspend pellet in 15 ml of Solution B and incubate on orbital shaker at 50 rpm in the dark at 25°C for 1–2 h.

6. Filter through sterile glass wool and rinse twice with 2–3 ml of Solution A.

7. To increase concentration of protoplasts, centrifuge at 600 *g* for 5 min and resuspend pellet in 5 ml of Solution A.

8. Use light microscope to examine preparation for intact protoplasts.

9. Assess small sample for viability using FDA (see *Protocol 7.3*).

Protocol 8.3

Rapid preparation of mesophyll protoplasts from maize leaves

Equipment

Razor blades

Forceps

Conical flask with a sidearm

Filter pump

Large Petri dish

Nylon sheet, 60-μm mesh (Henry Simon Ltd, P.O. Box 31, Stockport, Cheshire SK3 0RT, UK)

135-μm mesh spatula (Henry Simon Ltd)

Light microscope

Bench centrifuge.

Materials and reagents

Zea mays plants with well-developed leaves

Cellulase buffer (CB), 50 ml: 20 mM MES buffer pH 5.5 (0.43 g); 1 mM $MgCl_2$ (10.16 mg); 0.6 M sorbitol (8.2 g); 2% cellulase (1 g); 0.1% pectinase (0.05 g)

Wash buffer (WB), 100 ml: 50 mM Tris buffer (0.606 g) pH 8.0; 0.6 M sorbitol (16.4 g); 1 mM $MgCl_2$ (20.32 mg); 100 mM β-mercaptoethanol (0.701 ml).

Protocol

1. Take well-developed maize leaves and using the razor blade, slice into 0.5–1-mm strips.

2. Place strips in a vacuum flask containing 50 ml of cellulase buffer.

3. Apply vacuum using a filter pump. Maintain the negative pressure until all sections have become infiltrated with buffer.

4. Gently transfer to a large Petri dish and allow digestion to continue at room temperature for 3–5 h. Take a small volume (100 μl) to view under a light microscope to assess the level of digestion.

5. Filter the suspension through the 135-μm mesh to remove the broken cells and debris. Resuspend the partially digested fragments of leaf in 50 ml of WB in the original Petri dish.

6. Using the forceps, shake the leaf fragments in the WB to release the protoplasts. Agitating the Petri dish may also help.

7. Filter through the 60-μm mesh. Pellet the protoplasts by centrifugation at 300 **g** for 5 min. Gently resuspend in 25 ml of WB.

8. Repeat wash step and check viability of washed protoplasts with FDA (see *Protocol 7.3*).

Protocol 8.4

Preparation of mesophyll protoplasts from tobacco leaves and purification of protoplasts on a density gradient

---

Equipment

Razor blades

50 ml of digestion medium for 10–15 g of plant tissue

Plastic Petri dish, 9 cm

Glass wool

Nylon mesh 100–200 μm pore size (Henry Simon Ltd)

Bench centrifuge, capable of operating at very low g-forces and centrifuge tubes

Glass Pasteur pipette

Light microscope

Tea strainer

Fine forceps.

Materials and reagents

Solution A, 50 ml: Digestion medium 500 mM D-sorbitol 4.56 g; 1 mM $CaCl_2$ 7.35 mg; 5 mM MES-KOH, pH 5.5, 49 mg; Cellulase Onozuka R10, 1 g and Macerozyme R10, 0.15 g

Solution B, 100 ml: Wash medium 500 mM D-sorbitol 9.11 g; 1 mM $CaCl_2$, 14.7 mg and 5 mM MES-KOH pH 6.0, 98 mg

Solution C, 100 ml: density gradient solution 500 mM sucrose 8.56 g; 1 mM $CaCl_2$ 7.4 mg and 5 mM MES-KOH pH 6.0, 98 mg

Solution D, 100 ml: density gradient solution 400 mM sucrose 6.8 g; 100 mM D-sorbitol 0.90 g; 1 mM $CaCl_2$ 7.4 mg and 5 mM MES-KOH pH 6.0, 98 mg

Tobacco plant with well-developed leaves, 3–4 weeks old.

Protocol

1. Take a leaf and gently score underside with a sharp razor blade until the entire area is criss-crossed with fine, parallel cuts. Avoid cutting completely through the upper epidermis. Lay the cut area down on the surface of digestion medium in a Petri dish. Repeat this procedure until the surface of the Petri dish is covered.

2. Replace Petri dish lid and incubate in the dark for 16 h at room temperature. Repeat this, using fresh Petri dishes until there is sufficient material for your purposes.

3. Using fine forceps to hold the stem, shake the leaf on the surface of the digestion medium. The medium should change colour to green as proto-plasts leave leaf structure.

4. Collect the protoplast suspensions from each Petri dish and pool.

5. Pour through a tea strainer and then the nylon mesh to remove cell debris. Wash each membrane gently after use to remove protoplasts.

6. Centrifuge combined washes at 50–100 g for 3 min. The protoplasts should collect at the bottom of the centrifuge tube. Discard the supernatant.

7. Gently resuspend protoplast pellet in 10 ml of Solution C and aliquot 5 ml to each of two centrifuge tubes.

8. Slowly add 5 ml of Solution D to each centrifuge tube, then add 5 ml of Solution B to make a three step gradient. These solutions should be added gently while the tube is slanted so that they run slowly down the side of the tube and generates discrete steps.

9. Centrifuge at 300 g for 5 min.

10. The protoplasts will appear at the interface between the top two layers. Carefully remove the top layer with a pipette and then collect the protoplasts.

11. Check this preparation under the light microscope. Avoid putting a cover-slip on the slide as it can damage the fragile protoplasts. It should be possi-ble to see that the protoplasts are intact and not osmotically stressed. Stressed protoplasts can be identified by the cytoplasm appearing to be dense and compacted. Healthy protoplasts should also be perfectly spherical.

Protocol 8.5

Rapid preparation of protoplasts from maize roots

Equipment

Scalpel

Shaking water bath

Glass wool

Bench centrifuge capable of very low speeds and tubes

Parafilm

Petri dish

100-ml Erlenmeyer flask.

Materials and reagents

Protoplasting medium (PM), 50 ml: 5 mM MES-KOH (49 mg), pH 6.0; 0.5 M sorbitol (4.56 g), 1 mM $CaCl_2$ (7.35 mg); 0.5% BSA (0.25 g), 0.8% cellulase (Onozuka R10) (0.4 g), and 0.08% pectolyase (0.04 ml)

Wash medium (WM), 200 ml: 5 mM MES-KOH pH 6.0 (196 mg); 0.5 M sucrose (8.5 g) and 1 mM $CaCl_2$ (29.4 mg)

Density gradient step 1 (DG1), 50 ml: 5 mM MES-KOH (49 mg), pH 6.0; 0.1 M sorbitol (0.91 g); 0.4 M sucrose (3.4 g) and 1 mM $CaCl_2$ (7.35 mg)

Density gradient step 2 (DG2), 50 ml: 5 mM MES-KOH (49 mg), pH 6.0; 0.5 M sorbitol (4.56 g), 1 mM $CaCl_2$ (7.35 mg)

Maize seedlings grown hydroponically with 8–10 cm roots.

Protocol

1. Take roots and excise 8 cm from root tip, excluding any shoot material.

2. Using a sharp razor blade finely chop 20 g of root material in 20 ml of proto-plasting medium in a Petri dish. The tissue should be sliced to a fine pulp.

3. Transfer the homogenate to a 100-ml Erlenmeyer flask and incubate at 28°C in the dark for 3 h in a shaking water bath.

4. Filter the suspension through glass wool.

5. Centrifuge the filtrate at 60 *g* for 5 min.

6. Resuspend the pellet in 5 ml of ice-cold wash medium.

7. Layer on to this first 2 ml of DG1, then 1 ml of DG2 to form a step gradient (see *Protocol 8.4*, step 8).

8. Centrifuge this at 200 **g** for 5 min. The protoplasts will form a layer on the interface between the top two layers.

9. Wash protoplasts in 5 ml of DG2 and centrifuge at 100 **g** for 10 min. Resuspend pellet in 1 ml of DG2.

10. Assess for viability using FDA (see *Protocol 7.3*) and keep on ice until ready to use.

Protocol 8.6

Preparation of protoplasts from roots (arabidopsis) with partial purification of protoplasts by flotation

---

Equipment

Sterile 250-ml Erlenmeyer flask

0.22 μm filters to sterilize heat-labile solutions

Autoclave

pH meter

Sterile Petri dishes

Growth chamber with orbital shaker

Scalpel, scissors and forceps

Broad-mouthed pipette

100-, 50- and 25-μm sieves or meshes

Bench centrifuge capable of very low speeds and centrifuge tubes

Haemocytometer.

Materials and reagents

Sterile distilled water

10% v/v sodium hypochlorite containing 0.05 ml l^{-1} Tween 20

Basal medium (BM): 4.3 g l^{-1} of MS medium, 3% sucrose, B5 vitamins (Sigma), pH 5.8

0.5 BM agar plates (2.2 g l^{-1} MS, 3% sucrose, B5 vitamins, pH 5–8) 0.8% agar

50 ml aliquots of 0.5 BM in 250-ml Erlenmeyer flasks, sterile

MSAR I medium: BM plus 2 mg l^{-1} IAA, 0.5 mg l^{-1} 2,4-D, 0.5 mg l^{-1} 6-(γ,γ-dimethylallylamino)purine riboside (IPAR)

0.45 M sucrose solution

Protoplasting medium (PM): 0.5 BM containing 0.45 M sucrose or 0.45 M mannitol

Enzyme solution: 1% cellulase (Onozuka R10), 0.25% macerozyme (R10 Serva) dissolved in PM

0.45 M mannitol

Arabidopsis roots grown aseptically on agar (see Chapter 6 or 10).

Protocol

1. Grow arabidopsis seeds (var. Columbia or C 24) on 0.5 BM plates for 10–14 days in a light regime of 16 h light and 8 h dark at 22°C (see Chapter 6).

2. Aseptically remove plantlets from plate and separate the roots from the green tissue. Chop the roots into 2–4-mm pieces, using the sterile scalpel and transfer into Petri dishes with 10 ml of MSAR 1 medium.

3. Incubate in the dark or under low light conditions, shaking at 100 rpm for 7–12 days.

4. Remove the medium and wash the root explants once with 0.45 M sucrose solution. Replace with 20 ml of enzyme solution.

5. Incubate for 12–16 h with occasional shaking.

6. Collect the protoplasts with a broad-mouthed pipette and pass through the 100-, 50- and 25-μm sieves.

7. Centrifuge the suspension for 5 min at 50 rpm.

8. Collect the band of floating protoplasts concentrated at the top of the solution and transfer to a new tube. At this stage, it may be necessary to pool the contents of two or more tubes to obtain a working number of protoplasts for the next steps.

9. Resuspend in 0.45 M mannitol and pellet at 60 g for 5 min. Repeat this step twice to completely remove the enzymes.

10. Check for viability using FDA (see *Protocol 7.3*).

Protocol 8.7

The culture of maize protoplasts using nurse culture

Equipment

Autoclave

Tissue culture dishes (15 mm)

Balance

Sterile pipettes (5 ml) and pipetting device

Automatic pipette and sterile tips

Bench centrifuge capable of very low speeds and tubes

pH meter

Sterile 0.8-μm Millipore filters

Parafilm

Sterile forceps

Haemocytometer

Light microscope.

Materials and reagents

(Solutions should be sterilized)

Stock solution of 2,4-D at 1 mg/10 ml of ethanol : water 1:1

Stock solution of thiamine-HCl at 50 mg l^{-1} of H$_2$O (20 ml)

Modified White Vitamins: glycine 200 mg l^{-1}, nicotinic acid 50 mg l^{-1}, pyridoxine HCl 50 mg l^{-1}, thiamine HCl 50 mg l^{-1}

Protoplast culture medium (PCM) 100 ml (=4 × 9 cm Petri dishes): 0.43 g of MS salts, 2 ml of 2,4-D stock solution, 0.025 g of glucose, 1 ml of stock thiamine, 2 g of sucrose, 2 ml of coconut water, 3.64 g of mannitol, pH to 5.9 with 0.1 M KOH. PCMLMP: PCM medium with 0.8 g of low-melting-point (LMP) agarose (Gibco BRL/Life Technologies)

Postfeeder culture medium-MS 2D 100 ml: 0.43 g MS salts, 1 ml of Modified White Vitamins, 10 mg of myoinositol, 2 g of sucrose, 2 ml of stock 2,4-D, pH to 5.9 with 0.1 M KOH, 0.4 g of LMP agarose

Maize protoplasts (see *Protocols 8.3* or *8.5*).

Protocol

1. Prepare nurse cultures. These can be prepared during the protoplast digestion. The media should be prepared and autoclaved for 15 min at 121°C (see Chapter 5).

2. Aseptically pipette 3 ml of PCM-LMP into sterile, 60 × 15-mm culture dishes and allow to set.

3. Add an additional 3 ml of PCM-LMP containing 0.2 ml PCV of feeder cells from original cell suspension culture. PCM-LMP should be cooled to 40–45°C before mixing with the feeder cells.

4. Take 0.01 ml of the protoplast suspension and count. Adjust the concentration to 1×10^6 protoplasts ml^{-1} of PCM.

5. Plate 0.2 ml of the suspension on to the 0.8-µm filters resting on top of the feeder layer. Ensure that the plates are not too dry.

6. Leave the closed feeder plates in the laminar flow cabinet to let the protoplasts settle on to the filters. This may take from several hours to overnight.

7. Seal plates with Parafilm and incubate at 25°C in the dark.

8. Transfer filters carrying protoplasts to fresh feeder plates at weekly intervals using sterile forceps.

9. After 2–3 weeks, callus should be visible to the naked eye and can be moved using sterile tools to agar plates.

Protocol 8.8

The culture of arabidopsis protoplasts using the hanging drop method

Equipment

55-mm Petri dishes

Sterile wide-bore pipette

Sterile spatula

Growth chamber

Light microscope

Haemocytometer.

Materials and reagents

(Solutions should be sterilized)

Flask of freshly prepared protoplasts (see *Protocol 8.6*)

Basal medium (BM): 4.3 g l^{-1} of MS basal salts, B5 vitamins (Sigma)

Sodium alginate solution: 1% w/v solution of sodium alginate in BM medium containing 0.45 M sucrose

Calcium-agar plates: 20 mM CaCl$_2$, 0.45 M sucrose and 1% agar

PM, protoplast medium: 2.21 g l^{-1} MS basal salts containing 0.45 M sucrose, MSAR I hormones (2 mg l^{-1} IAA, 0.5 mg l^{-1} 2,4-D, 0.5 mg l^{-1} IPAR

MSAR I medium: 4.3 g l^{-1} MS plus 2 mg l^{-1} IAA, 0.5 mg l^{-1} 2,4-D, 0.5 mg l^{-1} IPAR.

Protocol

1. Suspend protoplast pellet from final wash in 1 ml of sodium alginate solution and count protoplast yield using haemocytometer. Adjust density to 5–7 × 10^5 cells ml^{-1} using more sodium alginate if necessary.

2. Using a wide-bore pipette, transfer the protoplast–sodium alginate suspension as 250–500-μl drops on the calcium agar plates.

3. Allow the alginate to form a gel for 45 min.

4. Transfer the individual droplets with a spatula into 55-mm Petri dishes containing 5 ml of PM and culture in a growth chamber at 22°C under dim light.

5. Remove 2.5 ml of PM and replace with 2.5 ml of fresh PM on days 7 and 14.

6. Remove 1 ml of PM and replace with 1 ml of MSAR I medium on days 21, 28 and 35.

7. By day 35, microcalli should be visible.

Note

This protocol can also be applied to protoplasts derived from roots and cell suspension cultures. However, for initiating liquid cultures of protoplasts, the density needs to be adjusted to 1×10^6 cells ml^{-1} until cell divisions start when it should be diluted with PM to 5×10^5 cells ml^{-1}.

Protocol 8.9

Detecting regeneration of cell wall in isolated protoplasts

―――――――――――――――――――――――――――――――――――――

Equipment

Epifluorescence microscope with BG 12 filter set

Slides and coverslips

Automatic pipetting device and tips.

Materials and reagents

Calcofluor White medium (CWM) 0.05% Calcofluor White in culture medium supplemented with 0.4 M sorbitol

Protoplast suspension.

Protocol

1. Resuspend a small volume of protoplasts in CWM medium and observe under an epifluorescence microscope. Calcofluor White stains only cell wall material and exhibits fluorescence when irradiated with blue light. As the cell wall is unlikely to reappear until the protoplasts have been in the appropriate regenerating medium for a minimum of 2 h, the Calcofluor White staining will not appear until this time has elapsed.

2. Initially, the staining will manifest itself as small, fluorescent dots on the surface of a small percentage of the population. After 5–6 h has elapsed, more protoplasts will show this pattern of staining.

3. If the treatment is continued for 18–20 h, large areas of dense staining will appear on viable, healthy protoplasts. Eventually, the entire cell surface will be covered with fluorescent staining as the cell wall is regenerated completely.

Protocol 8.10

Protoplast fusion induced by polyethylene glycol (PEG)

Equipment

Bench centrifuge capable of very low speed (50 **g**) and tubes

Pipette

Timer

Parafilm

Petri dishes (9 cm).

Materials and reagents

(Solutions should be sterilized)

PEG fusion solution: 30% w/v PEG 6000, 4% w/v sucrose, 10 mM $CaCl_2$. This should be autoclaved and stored in dark until used.

Protoplast maintenance medium appropriate to the types of protoplasts to be used. Also to be used as a washing medium. These should be sterile.

9-cm Petri dishes containing 8 ml of sterile protoplast medium plus 0.8% agarose.

Protocol

1. Ensure that the densities of the two protoplast populations are both 2×10^5 cells ml^{-1}.

2. Spin 4 ml of each population together in one tube, at 100 **g** for 10 min to pellet the protoplasts.

3. Remove the surplus medium so as to leave the pellet in 0.5 ml of medium.

4. Add 2 ml of PEG fusion solution to each tube and leave for 10 min.

5. Sequentially dilute the fusion mixture at 5-min intervals, by adding 0.5, 1, 2, 2, 3 and finally 4 ml of protoplast maintenance medium per tube.

6. Gently mix tube after each addition.

7. Pellet protoplasts at 100 **g** for 10 min and remove supernatant.

8. Wash the protoplasts once in protoplast medium.

9. Resuspend finally in 16 ml of protoplast medium.

10. Aliquot 8 ml, using a wide-mouthed pipette, onto the surface of the agarose plates.

11. Seal the plates with Parafilm and culture at 25°C under a low light regime.

12. Microcalli should be visible after several weeks. Blocks of agarose may be aseptically cut out and transferred to liquid culture if required.

Protocol 8.11

Electrical fusion of protoplasts

Equipment

Electrofusion apparatus including sterile fusion chambers and electrodes.

Materials and reagents

(These media should be sterile)

Protoplast maintenance solutions suitable for both populations

Solid media suitable for post-electrofusion protoplasts, details of precise media for the species chosen should be collected from the literature.

Protocol Part A – Electrofusion

1. The suspension of mixed protoplast populations should be placed in the electrofusion chamber. A weak AC current (400 000 Hz, 1.5 V) is passed between the electrodes, through the protoplasts, causing them to become positively charged on one side and negatively charged on the other. The protoplasts align themselves in chains. Varying protoplast density, frequency of AC field and peak-to-peak voltage can alter the numbers in each chain.

2. After a period of approximately 90 s, the AC field is replaced by a single, high-voltage pulse discharge of, for example, 1000 V cm^{-1}. Where the protoplasts are in contact with each other the plasma membranes break and fuse forming a continuous membrane around the pearl chains. This is followed by cytoplasmic fusion. Multiple fusions will occur at high concentrations of protoplasts but this can be adjusted with experience.

3. The protoplast suspension is then plated out and cultured as in Protocol 8.10 or as in the literature for the species used. Electrofusion can be less productive than PEG fusion.

Protocol Part B – Selection of heterokaryons

After chemical or electrical fusion of protoplasts, there will be a mixture of unfused protoplasts, fused parents of the same origin, multiple fusions and the desired heterokaryons. Isolation of the correct product can be achieved by a choice of the methods below.

1. The simplest technique is to allow all products of fusion to mature and regenerate and then to identify the heterokaryons by morphological differences at the seedling stage.

2. If the two parent populations have obviously different visual properties, it may be possible to separate fused heterokaryons using a microscope and a micromanipulator. This procedure may be laborious in practice.

3. If the parents are not easily distinguishable by sight, the parent populations may be stained with two different colour dyes. For example, FDA (see *Protocol 7.3*) fluoresces yellow–green while rhodamine isothiocyanate is red. Selection of fused protoplasts showing dual fluorescence allows identification of heterokaryons.

4. A flow cytometer equipped with a UV source can be calibrated to separate dual-labelled protoplasts from single-labelled. Flow cytometry can process a large number of protoplasts accurately.

Protocol 8.12

Protoplast transformation by electroporation

Equipment

All equipment required to prepare protoplasts (see previous protocols)

Small Petri dishes

Ice and ice bucket

Electroporation apparatus with 1.0 ml cuvettes containing built-in electrodes

Timer.

Materials and reagents

(Solutions should be sterilized)

Electroporation buffer (EB): 1 mM MES, pH 5.8, 2.5 mM $CaCl_2$, 0.225 M sorbitol, 0.225 M mannitol

MS medium: appropriate culture medium supplemented with 0.6 M sorbitol.

Protocol

1. Resuspend protoplast pellet gently in 25 ml of EB. Spin at 600 *g* for 10 min.

2. Repeat protoplast wash 4–5 times with 25–50 ml of EB.

3. Resuspend final pellet in 5 ml of EB.

4. Add DNA in a ratio of 100 μl of DNA : 500 μl cells.

5. Aliquot 600-μl volumes into Petri dishes for electroporation and leave on ice for 5 min.

6. Electroporate, following manufacturer's instructions and recommendations. Typical conditions recommended are: capacitance 910 μF, voltage 120–140 V, time of discharge 1000 ms, volume 600 μl and electrode distance is 0.2 mm.

7. Rest the electroporated protoplasts on ice for 15 min.

8. Collect the electroporated protoplasts in a Petri dish. Add 0.9 ml of EB and 1.6 ml of MS medium.

9. Incubate in dark at 22°C. Expression should be apparent after 12 h.

Protocol 8.13

Transformation of protoplasts mediated by polyethylene glycol (PEG)

Equipment

Ice and ice bucket

Bench centrifuge capable of low speed and tubes

Haemocytometer and microscope

Glass Petri dish (9 cm)

Automatic pipetting device and tips.

Materials and reagents

All solutions needed to prepare protoplasts (see previous protocols). These need to be sterile

0.5 M MaMg solution; 0.5 M mannitol, 15 mM $MgCl_2.6H_2O$, 0.2% MES, pH 5.8 (sterile)

PEG solution: 40% w/v PEG 1450 in MaMg solution. This should be filter-sterilized

Plasmid DNA solution in water.

Protocol

1. Resuspend protoplast pellet in 1 ml of MaMg solution.

2. Count cells and adjust density to approximately 1×10^6 cells ml^{-1}.

3. Place protoplast suspension on ice for 35 min.

4. Centrifuge protoplasts at 60 g for 5 min.

5. Resuspend pellet in 0.3 ml of MaMg solution and carefully transfer them as single droplets to the middle of a 9-cm glass Petri dish.

6. Slowly add 25–35 µg of plasmid DNA dissolved in water into the drop of protoplast suspension. Rock the Petri dish very gently to mix the protoplasts and the DNA.

7. After 5 min, add 0.3 ml of PEG solution at the circumference of the drop of protoplast suspension.

8. Again tilt the Petri dish very gently to mix the PEG and the protoplasts. Alternatively, mix with a sterile micropipette tip.

9. After 10 min, add 1 ml of 0.45 M mannitol solution to the sides of the droplet.

10. At 2-min intervals, add 2 ml of 0.45 M mannitol solution with gentle shaking until about 12 ml are present in the Petri dish.

11. Collect the suspension by wide-mouthed pipette and transfer to a centrifuge tube. Centrifuge at 60 g for 5 min.

12. At this stage, the protoplasts can be studied as a transient expression system or transferred to selective culture medium to regenerate stably transformed calli or plantlets. If the protoplasts are to be subcultured, the entire procedure should be carried out under aseptic conditions.

Chapter 9

Haploid cultures

1. Introduction

One of the most significant biotechnological advances in commercial plant breeding in recent years has been the development of *in vitro* techniques to produce haploid plants that can be used to generate homozygous lines.

In traditional plant breeding programmes, crosses between distantly related species can bring about novel gene combinations. However, the hybrid offspring can be few in number and may be genetically unstable. More importantly, these hybrids require several generations (6–10) of further crossing or selfing to attain the degree of homozygosity that is necessary to ensure that advantageous characteristics are fixed and to eliminate any undesirable recessive traits. Alternative procedures for the production of homozygous lines in one or two generations can therefore save several years of conventional plant breeding. The formation of dihaploids from haploid plants is a quick and valuable way of producing homozygous lines from heterozygous parents.

Haploid plants have a simple or gametophytic set (1n) of chromosomes and may be sterile. However haploid plants can double their chromosome set spontaneously or this can be induced by chemical agents (e.g. colchicine, the mitotic spindle disrupter, results in chromosome duplication without cell division) to produce fertile doubled haploids (dihaploids). An advantage of this procedure is that the resulting dihaploids are homozygous for all genes and therefore true-breeding, making the selection process for desirable traits more efficient.

In theory, haploid plants could be produced by the *in vitro* culture of haploid gametophytic tissues, that is, unpollinated ovaries or ovules and anthers or pollen. Viable haploid plants have been produced by ovary and ovule culture for some crops; the procedures are however technically laborious and the yields low. There are two *in vitro* procedures that are now routinely used in the production of haploids: (i) anther culture and (ii) the culture of isolated microspores (see *Figures 9.1* and *9.2*). The development of haploid plants from male gametophytic tissue is known as androgenesis.

2. Anther culture

Anther culture (*Protocol 9.1*) is a rapid and relatively simple procedure by which pollen-producing anthers are collected from flowers at a certain stage of development; sometimes they are given a low- or high-temperature pre-treatment, before sterilization and culturing on a specific nutrient medium. Mature pollen (the male gametophyte) develops from haploid microspores that are in turn formed by meiosis in pollen mother cells. Microspores have

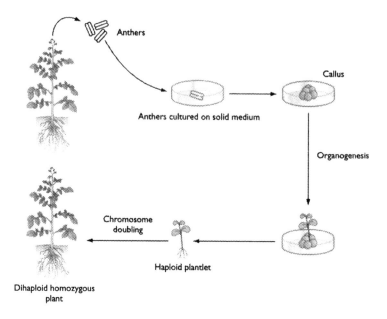

Figure 9.1

Outline of anther culture. Dihaploids are formed by treatment of haploid tissue with colchicine or other chemical agents.

a single nucleus and mature pollen three nuclei. When anthers containing microspores at the middle to late uninucleate stage are cultured on solid media, individual microspores form callus. Plants with half the chromosome number can be regenerated from the callus either directly by embryogenesis or indirectly by organogenesis. Alternatively, microspores may give rise to embryoids (aggregates of cells with recognizable embryonic structures) that are capable of differentiation into embryos (*Figure 9.1*).

3. Microspore culture

Microspore culture (*Protocol 9.2*) involves the physical isolation of microspores from anthers at a specific stage of development. After a suitable pre-treatment, the anthers are homogenized and microspores are isolated and purified from the homogenate by filtration and centrifugation steps. The purified microspores are then cultured in a liquid medium. Small embryos develop directly from the microspores (microspore embryogenesis) and are transferred to a regeneration medium for plantlet development (*Figure 9.2*).

4. Anther culture versus microspore culture

Androgenesis has been used to produce haploids from several hundred different plant species including commercially important cultivars. The choice of which procedure to adopt for haploid production is determined by a number of considerations. Anther culture is a relatively simple and rapid technique that can produce large numbers of haploids with minimal equipment. The major problems generally encountered in the use of anther culture are genotype dependency and the high frequency of albino plants and mixed ploidy types (mixoploids) in the regenerated populations. In

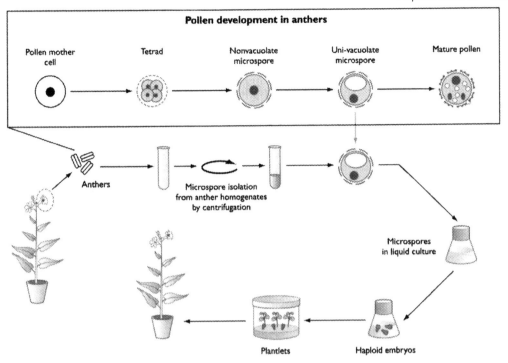

Pollen development in anthers

Pollen mother cell — Tetrad — Nonvacuolate microspore — Uni-vacuolate microspore — Mature pollen

Anthers

Microspore isolation from anther homogenates by centrifugation

Microspores in liquid culture

Plantlets

Haploid embryos

Figure 9.2

Isolation and culture of microspores.

comparison to anther culture, the microspore technique is technically more complicated and requires more extensive laboratory facilities. However, in species where experimental comparisons have been made, problems such as mixoploidy are less frequent in populations derived from microspore cultures than anther cultures. The use of microspore cultures also has other advantages: (i) development of embryos directly from microspores without an intermediate callus stage reduces the risk of genetic abnormality; (ii) microspore cultures are ideal material for genetic transformation, as insertion of foreign DNA into the haploid genome will give rise to homozygous expression after chromosome doubling to form the dihaploid.

5. Intergeneric crosses and embryo rescue

An alternative method for rapid haploid production involves intergeneric crossing followed by embryo rescue. Such wide crosses may be incompatible and viable seed development may not occur. However, a haploid embryo may form and this can be rescued by excision and tissue culture. The basis of this procedure relies on fertilization to produce a diploid zygote, which is then followed by elimination of one set of chromosomes from the developing embryo to produce a haploid embryo. To ensure its survival the embryo must be dissected out and transferred to nutrient culture media. This technique has been successful in cereals such as barley and wheat. For example in the so-called *bulbosum* technique, bread wheat *Triticum aestivum* (female parent) is crossed with the barley *Hordeum bulbosum* to yield haploid plants.

Other wide crosses include bread wheat and durum wheat *Triticum turgidum* spp. *durum* with maize *Zea mays* as the paternal parent.

6. Procedures for the induction of androgenesis

Although androgenesis in cereals was noticed more than 70 years ago, the development of protocols to exploit this phenomenon in plant breeding has only been achieved recently. Androgenesis has been tested in several economically important plants, but many crops are still recalcitrant. As yet, there is no universal protocol common to all plant species. In general, protocols consist of three sequential phases. The protocol for barley microspore embryogenesis is a typical example. In this case, the procedure is divided into three principal steps:

(i) Pre-treatment of anthers – anthers at the right developmental stage are given a stress pre-treatment for approximately 4 days.
(ii) Microspore culture – microspores are isolated from the treated anthers and put into culture. Cell division can be observed after 4 days and multicellular structures are formed after 14 days.
(iii) Development of embryos – multicellular structures develop into embryos and further growth into plantlets requires an additional 21 days.

There are many variations of this basic protocol. Successful microspore regeneration depends on the plant material. The variety, age and the growth status of the plant are all important factors. The role of the pretreatment stress is to induce the microspores to switch from the gametophytic developmental pathway to sporophytic development. The pretreatment may be temperature (cold or heat shock), starvation or osmotic stress, and can be applied to intact flowers (e.g. barley spikes), isolated anthers or isolated microspores. The stress hormone abscisic acid is implicated in the reprogramming of microspore development.

During the pre-treatment, some of the microspores develop into embryogenic microspores that can develop into embryos. Embryogenic microspores are characteristically twice the size of non-embryogenic microspores. After pre-treatment, the microspores are cultured in a specific medium where cell division and differentiation occur. At this stage, chromosome doubling to form dihaploids may occur spontaneously or can be induced by agents such as colchicine.

Further reading

Holme, I.B., Olesen, A., Hansen, N.J.P. and Andersen, B. (1999) Anther and isolated microspore culture response of wheat lines from northwestern and eastern Europe. *Plant Breeding* 118: 111–117.
Wang, W., van Bergen, S. and Van Duijn, B. (2000) Insights into a key developmental switch and its importance for efficient plant breeding. *Plant Physiology* 124: 523–530.

References

Chu, C.C., Hill, R.D., Brule-Babel, A.L. (1990) High frequency of pollen embryoid formation and plant regeneration in *Triticum aestivum* L. on monosaccharide containing media. *Plant Sci.* 66: 255–262.
Wang, X. & Hu, H. (1984) The effect of potato II medium for triticale anther culture. *Plant Sci. Lett.* 36: 237–239.

Protocol 9.1

nther culture of wheat (*Triticum aestivum* L.)

Equipment

Sterile dissecting instruments

Dissecting microscope

Incubator, 28°C.

Materials and reagents

Greenhouse grown wheat plants

Petri dishes (60 × 14-mm) containing 4.5 ml of CHB-2 medium [KNO_3, 1415 mg l^{-1}; $(NH_4)_2SO_4$, 232 mg l^{-1}, KH_2PO_4, 200 mg l^{-1}; $CaCl_2.2H_2O$, 83 mg l^{-1}; $MgSO_4.7H_2O$, 93 mg l^{-1}, $FeNa_2$ EDTA, 32 mg l^{-1}; $ZnSO_4.7H_2O$, 5 mg l^{-1}; $MnSO_4.4H_2O$, 5 mg l^{-1}; H_3BO_3, 5 mg l^{-1}; KI, 0.4 mg l^{-1}; $Na_2MoO_4.2H_2O$, 0.0125 mg l^{-1}; $CuSO_4.H_2O$, 0.0125 mg l^{-1}; $CoCl_26H_2O$, 0.0125 mg l^{-1}; glycine 1.0 mg l^{-1}; thiamine-HCl, 2.5 mg l^{-1}; Pyridoxine-HCl, 0.5 mg l^{-1}, nicotinic acid, 0.5 mg l^{-1}; biotin, 0.25 mg l^{-1}; calcium pantothenate, 0.25 mg l^{-1}; ascorbic acid, 0.5 mg l^{-1}; myo-inositol, 300 mg l^{-1}; glutamine, 1000 mg l^{-1}; glucose, 0.21 mg l^{-1}; 2,4-D, 0.5 mg l^{-1}; kinetin, 0.5 mg l^{-1}; pH 4.5, supplemented with 9% (w/v) maltose (sterile); Chu et al., 1990]

70% Ethanol spray

Sodium hypochlorite (20%)

Acetocarmine solution (commercial preparation)

Sterile distilled water

Sterile filter paper.

Protocol

1. Check microspore development in anthers from spikes just prior to emergence of first leaf blade. Remove anthers, stain with acetocarmine and examine with a microscope.

2. Collect spikes containing anthers with microspores in the mid- to late-uninucleate stage.

3. Store spikes with their cut ends in tap water at 5°C for 5 days.

4. Spray spikes with 70% alcohol, and surface sterilize with sodium hypochlorite for 20 min.

5. Wash spikes three times with sterile water and blot dry with sterile paper tissue.

6.   Dissect out the anthers under sterile conditions and add anthers from one spike to each Petri dish containing CHB-2 medium.

7.   Incubate cultures in the dark at 28°C for 35 days.

8.   For plant regeneration – at the end of the incubation period, remove embryos and transfer to 190-2 medium (Wang and Hu, 1984) solidified with 3.5 g l^{-1} Gelrite. Incubate at 25°C and with a light intensity of 60–80 μmol m^{-2} s^{-1}.

Protocol 9.2

Wheat microspore culture

Equipment

Blender (Waring or similar, container sterilized)

Stainless steel mesh 100-μm (sterilized)

Bench centrifuge with swing-out rotor and sterile centrifuge tubes

Haemocytometer

Incubator 28°C

Microscope (ultraviolet).

Materials and reagents

Materials and reagents as for *Protocol 9.1*

Mannitol 0.3 M (sterile)

Maltose (21% w/v) (sterile)

Petri dishes (35 × 10-mm)

Fluorescein diacetate (*Protocol 7.3*).

Protocol

1. Grow plants and collect and treat spikes as described in *Protocol 9.1* (steps 1–5).

2. Remove spikelets from five spikes and blend with a Waring blender in 0.3 M mannitol.

3. Filter slurry through the stainless steel mesh.

4. Centrifuge filtrate at 55 *g* for 5 min to pellet microspores.

5. Resuspend microspore pellet in 0.3 M mannitol.

6. Layer suspension on top of a 21% (w/v) maltose solution in a centrifuge tube.

7. Centrifuge at 55 *g* for 5 min. Collect microspores accumulated at gradient interface.

8. Wash microspore fraction once by resuspension in mannitol and pelleting at 55 *g* for 5 min.

9. Check viability of microspores with FDA and determine density with a haemocytometer.

10. Resuspend in CHB-2 medium with 9% (w/v) maltose at a concentration of 50 000 microspores ml^{-1}.

11. Transfer 2.0 ml of microspore suspension to 35 × 10-mm Petri dishes and incubate at 28°C for 28 days.

12. For plant regeneration – at the end of the incubation period, remove embryos and transfer to 190-2 medium (Wang and Hu, 1984) solidified with 3.5 g l^{-1} Gelrite. Incubate at 25°C and with a light intensity of 60–80 μmol m^{-2} s^{-1}.

Chapter 10

Organ and embryo culture

1. Introduction

Maintaining and growing plant organs in culture has a variety of applications, from the experimental understanding of plant development to their use for the production of high-value products. An important feature of organ culture is that the organ grows and develops in a manner similar to that of the parent plant; this distinguishes them from callus or suspension-cultures (cell and tissue cultures) described in other chapters of this book.

A wide range of plant organs has been successfully cultured. The earliest experiments undertaken by White (1934) showed that root explants cultured in a liquid medium with inorganic nutrients, sucrose and yeast extract continued to grow at a rate of about 10 mm day^{-1}. Lateral roots developed after 4 days and could be used to initiate new root cultures. Protocols for root cultures (*Figures 10.1* and *10.2*) will be described later in this chapter. Subsequently, leaves, shoot tips, complete flowers, anthers, pollen, ovules and intact fruit have all been successfully cultured. A protocol for maize ear culture (based on Greyson, 1994) is given as an example of the culture of an entire organ.

2. Hairy roots

Hairy root cultures are generated when a plant tissue is transformed with a culture of the bacterium *Agrobacterium rhizogenes*. The method of transformation is similar to that of *A. tumefaciens*, which transfers a segment of plasmid DNA to the cells of an infected plant (Chapter 4). The transformation results in altered hormone levels, causing the induction of roots and a root system that branches much more frequently than the usual root system of that plant, and is covered with a mass of tiny root hairs. This root system may be maintained indefinitely in liquid culture without the parent plant. Importantly for cell and tissue culture, the transformed roots produce secondary metabolites at levels similar to those of the original plant. Hairy root cultures are a specialized form of organ culture and can be maintained continuously in bioreactors (Chapter 14). *Protocol 10.3* is given for the production of a hairy root culture of the dicot tomato. It is based on the method of Jones and Sutton (1997), and could be adapted for hairy root cultures of other dicots.

3. Embryo culture

Embryo culture, where the embryo is derived from the seed of a parent plant, is an important method both for the study of seed and embryo development, and as a means of obtaining viable seedlings when germination is a

Figure 10.1

Roots of Arabidopsis *grown in suspension culture in a conical flask.*

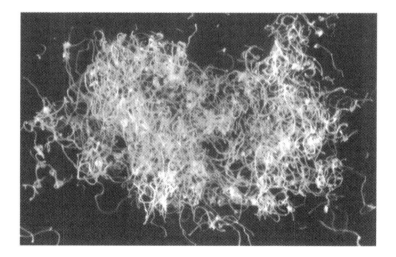

Figure 10.2

An enlarged view of roots of Arabidopsis *grown in suspension culture.*

problem. The easiest method of culturing an embryo is to remove it from the seed when nearly mature and ready to germinate (*Protocol 10.5*). However, in other instances (for instance to obtain sterile seedlings or to rescue mutant embryos), it may be necessary to dissect out an immature embryo and grow it directly on solid medium. The nutrient requirements of embryos that have been removed from the ovary is much more complex. Supplementing the culture medium with coconut water, a plant liquid endosperm (Chapter 5) was shown by van Overbeek in 1941 to be successful in providing the nutrients and growth factors required. As embryos mature, they become increasingly autotrophic, and more complex media are not required. A variety of media have been used for embryo culture. In addition to a basal medium (e.g. Gamborg's salts), sucrose or glucose is the usual carbohydrate source; the amino acid glutamine is preferred as nitrogen source in comparison with asparagine (Raghavan, 1980). Other additions include hormones (particularly auxins and cytokinins) and undefined additions like coconut milk and casein hydrolysate. The concentration of agar has been found to be of particular importance as too high a concentration leads to dehydration of the embryo.

Embryo rescue techniques are important where a high rate of spontaneous abortion of embryos is likely to occur, for instance in the production of hybrids between closely related species, or in the production of transgenic plants. In the technique, embryos are isolated and grown on solid media as described above for embryo culture. The method described below for embryo rescue is for interspecific hybrids of *Phaseolus vulgaris* and *Phaseolus polyanthus* (Geerts et al., 1999). It involves the removal of immature (early heart-shaped) embryos from seeds and their culture on defined media based on those of Gamborg or MS (see Chapter 5), supplemented with vitamins, a nitrogen source, the cytokinin BAP with sucrose as carbon source and 0.8% agar as gelling agent. After 20 days, the embryos are transferred to a rooting medium (with GA_3) before being established after a further 15 days in the glasshouse.

4. Minitubers and microtubers

Organ culture techniques are important in commercial potato growing where disease-free tubers are required for propagation. Minitubers are produced *in vitro* by plantlets from meristem tip culture in sterile culture. Plantlets develop in 4–8 weeks from a meristem explant and are routinely subcultured by nodal cuttings. Rooted cuttings are then used to generate minitubers – small seed tubers, less than 3 cm in diameter in sterile media. The plants that produce the minitubers are tested extensively for major potato pathogens. Minitubers are stored in a controlled atmosphere until needed, when sprouting is induced. Alternatively, microtubers, which are up to 5 mm diameter, are produced in *in vitro* culture in the axils of the leaves. Microtubers can be either planted in the greenhouse to produce plantlets or minitubers, or directly planted into the field.

References

Geerts, P., Mergeai, G. and Baudoin, J.-P. (1999) Rescue of early heart shaped embryos and plant regeneration of *Phaseolus polyanthus* Greenm. and *Phaseolus vulgaris* L. *Biotechnol. Agron. Soc. Environ.* **3**: 141–148.

Greyson, R.I. (1994) Maize inflorescence culture. In: The Maize Handbook (eds. M. Freeling and V. Walbot). Springer-Verlag, Berlin.

Jones, P.G. and Sutton, J.M. (1997) *Plant Molecular Biology – Essential Techniques*. John Wiley & Sons, Chichester.

Raghavan, V. (1980) Embryo culture. *Int. Rev. Cytol. Suppl.* **11**B: 209–240.

White, P.R. (1934) Unlimited growth of excised tomato root tips in a liquid medium. *Plant Physiol.* **9**: 585–600.

Protocol 10.1

Isolation and culture of the primary seedling root of dicots

Equipment

3 × 250-ml conical flasks containing 200 ml of sterile distilled water

24 × 90-mm Petri dishes

12 × glass culture tubes (150 × 25-mm) containing 20 ml of sterile distilled water

12 × glass culture tubes (150 × 25-mm) containing 20 ml of rooting medium (MR) culture medium

3 × scalpels

2 × pairs forceps.

Materials and reagents

MR: $Ca(NO_3)_2.4H_2O$, 236 mg l^{-1}; $FeSO_4.7H_2O$, 2 mg l^{-1}; KH_2PO_4, 12 mg l^{-1}; KNO_3, 81 mg l^{-1}; KCl, 65 mg l^{-1}; $MgSO_4.7H_2O$, 36 mg l^{-1}; sucrose, 4%; pH 5.8

24 pea seeds

250-ml conical flask containing 200 ml of 10% w/v calcium hypochlorite

1 × Pyrex conical flask containing 100 ml of 95% ethanol

2 × pieces of muslin (60 × 60-mm)

1 × 250-ml conical flask containing 200 ml of 1% aq. Tween 20.

Protocol

1. Wrap seeds in muslin and immerse in 95% ethanol for 10 s, 1% aq. Tween 20 for 1 min, 10% w/v calcium hypochlorite for 20 min. Proceed in aseptic conditions using a laminar flow hood.

2. Rinse the seeds three times in distilled water.

3. Put two seeds into distilled water in each of eight Petri dishes.

4. Germinate in the dark for 48 h.

5. Excise the apical 10 mm of the primary root and place into 20 ml of MR medium in a culture tube.

6. Incubate in the dark at 25°C. Lateral roots will develop after 5–6 days.

7. Remove and reculture lateral root tips.

Protocol 10.2

Isolation and culture of roots of monocots (e.g. maize)

Equipment

Orbital shaker

24×90-mm sterile Petri dishes

$2 \times$ pairs forceps (sterile)

Filter paper (sterile)

Sterile culture tubes (20 ml).

Materials and reagents

White's basal salts – available commercially or make: H_3BO_3, 1.5 mg l^{-1}; $Ca(NO_3)_2.4H_2O$, 200 mg l^{-1}; ferric sulphate, 2.5 mg l^{-1}; $MgSO_4.7H_2O$, 360 mg l^{-1}; $MnSO_4.H_2O$, 5 mg l^{-1}; KCl, 65 mg l^{-1}; KI, 0.75 mg l^{-1}; KNO_3, 80 mg l^{-1}; KH_2PO_4, 12 mg l^{-1}; $NaH_2PO_4.H_2O$, 16.5 mg l^{-1}; Na_2SO_4, 200 mg l^{-1}; sucrose, 2–4%; pH 5.8

3×250-ml conical flasks containing 200 ml of sterile distilled water

5×250-ml conical flasks containing 50 ml of White's medium (sterile)

24 maize kernels

250-ml conical flask containing 200 ml of 5% w/v sodium hypochlorite.

Protocol

1. Immerse kernels in 5% w/v sodium hypochlorite for 20 min. Proceed in aseptic conditions using a laminar flow hood.

2. Rinse kernels $3 \times$ in sterile distilled water (culture tubes).

3. Put two kernels onto sterile filter paper wetted with sterile distilled water in each of eight Petri dishes.

4. Germinate in the dark at room temperature.

5. Excise the apical 10 mm of the primary root and place into 50 ml of White's medium in a 250-ml conical flask.

6. Incubate in the dark at 25°C on an orbital platform at 50–60 rpm. Lateral roots will develop after 5–6 days.

7. Remove and reculture lateral root tips.

Protocol 10.3

Hairy root cultures (plant transformation with *Agrobacterium rhizogenes*)

Equipment

Filter paper (Whatman no. 1) 90-mm (sterile)

Scalpels (sterile)

Syringe and needles (sterile)

Forceps (sterile)

Conical flasks (250-ml, sterile)

Incubator for *A. rhizogenes* culture (28°C, see Chapter 13)

Incubator for Petri dishes – 22°C

Orbital incubator – 16°C dark, 120 rpm.

Materials and reagents

MS basal medium (sterile) (Chapter 5)

Petri dishes (90-mm) of 1/2 × MS basal medium + 3% sucrose and 0.2% Phytagel with Gamborg's B5 vitamins: 1 mg l^{-1} nicotinic acid, 1 mg l^{-1} pyridoxine HCl, 10 mg l^{-1} thiamine HCl, 100 mg l^{-1} myoinositol (sterile)

Petri dishes (90-mm) of 1/2 × MS basal medium + 3% sucrose and 0.2% Phytagel with Gamborg's B5 vitamins (sterile) and 10 μg ml^{-1} kanamycin and 250 μg ml^{-1} cefotaxime

Petri dishes containing yeast mannitol agar (YMA) medium: 2 mg l^{-1} MgSO$_4$, 0.5 mg l^{-1} K$_2$HPO$_4$, 0.1 mg l^{-1} NaCl, 10 mg l^{-1} mannitol, 0.4 mg l^{-1} yeast extract, 8 g l^{-1} agar (sterile) with 50 μg ml^{-1} rifampicin and 50 μg ml^{-1} neomycin added after autoclaving.

Antibiotic stocks (stored at −20°C until used)

Rifampicin (50 mg ml^{-1} in methanol)

Neomycin (50 mg ml^{-1} in distilled water)

Cefotaxime (250 mg ml^{-1} in distilled water)

Kanamycin (10 mg ml^{-1} in distilled water).

Agrobacterium

Agrobacterium rhizogenes (e.g. strain LBA9402, Rif®) with binary vector.

Protocol

Proceed in aseptic conditions using a laminar flow hood:

1. Surface-sterilize seeds and germinate in a sterile Petri dish (see Chapter 6).

2. Take an overnight culture of A. *rhizogenes* grown on YMA medium plates with 50 µg ml^{-1} rifampicin and 50 µg ml^{-1} neomycin at 28°C.

3. Take the seedlings and cut off the roots. Stab a sterile syringe needle first into the A. *rhizogenes* culture and then into the hypocotyl. Repeat for multiple inoculation sites.

4. Transfer the seedlings to Petri dishes containing 1/2 MS medium with B5 vitamins.

5. Incubate for 48 h at 22°C, 16 h light.

6. Incubate at 22°C in the dark until roots form (7–10 days).

7. Transfer to 1/2 MS plates with B5 vitamins and 10 µg ml^{-1} kanamycin and 250 µg ml^{-1} cefotaxime.

8. The roots should be subcultured to fresh plates when required.

9. To establish a liquid culture, excise the apical 10 mm of the root and transfer to a 250-ml conical flask containing 50 ml of 1/2 MS medium with B5 vitamins. Incubate on an orbital incubator at 120 rpm at 16°C and subculture to fresh medium when the roots begin to become brown.

Protocol 10.4

terile culture of the ear of a cereal (e.g. maize)

Equipment

Dissecting needles (sterile)

Scalpel (sterile)

Forceps (sterile)

Aluminium foil caps (sterile)

Incubator at 28°C.

Materials

(Sterile)

Dispensed into 125-ml conical flask (40 ml per flask): MS basal medium (*Protocol 5.1*) with White's vitamins and glycine (folic acid, 0.5 mg l^{-1}; nicotinic acid, 0.1 mg l^{-1}; thiamine HCl, 1.0 mg l^{-1}; pyridoxine HCl, 0.1 mg l^{-1}) glycine, 3 mg l^{-1}; I-inositol, 100 mg l^{-1}; $FeSO_4.7H_2O$, 27.8 mg l^{-1}; Na_2EDTA, 37.3 mg l^{-1}; sucrose, 6% w/v; kinetin, 10^{-7} M

3 × 250-ml beakers containing distilled water covered in foil

1 × 250-ml beaker containing 10% sodium hypochlorite.

Protocol

1. Take ears when 5–10 mm long.

2. Surface sterilize in 10% sodium hypochlorite (15 min).

3. Transfer to a laminar flow cabinet.

4. Rinse 3 × with distilled water.

5. Place ear into medium in conical flask.

6. Cover with sterile aluminium foil cap.

7. Incubate at 28°C, 18 h light, 6 h dark.

Protocol 10.5

Embryo rescue of a dicot – *Phaseolus vulgaris*

Equipment

Mounted needles (2, sterile)

Pasteur pipettes (10, sterile)

Scalpel (sterile)

Dissecting microscope

Incubator at 26°C.

Materials and reagents

(Sterile)

Ovule suspension medium Sucrose, 120 mg l^{-1}; agar, 0.75 g l^{-1} in distilled water

G1 medium in 90-mm plastic Petri dishes: Gamborg salts (Chapter 5), NH$_4$NO$_3$ 400 mg l^{-1}; sucrose 30 mg l^{-1}; thiamine HCl 1 mg l^{-1}; nicotinic acid 5 mg l^{-1}; pyridoxine HCl 0.5 mg l^{-1}; myo-inositol 100 mg l^{-1}; casein hydrolysate 1 mg l^{-1}; L-glutamine 1 mg l^{-1}; BAP 0.028 mg l^{-1}; Difco agar 8 g l^{-1}

R1 as 20-ml aliquots in 20 × 150-mm culture tubes: as G1, but without NH$_4$NO$_3$ and with L-glutamine 100 mg l^{-1}; casein hydrolysate 100 mg l^{-1}; GA$_3$ 0.03 mg l^{-1}

Sterile distilled water (3 × 200 ml) in 250-ml conical flasks

Ethanol (200 ml 70% w/v) in a 250-ml conical flask

200 ml of calcium hypochlorite in a 250-ml conical flask.

Protocol

1.   Harvest young seed pods 9–11 days after pollination.

2.   Surface sterilize the pods in 70% ethanol for 1 min.

3.   Immerse pods in 5% calcium hypochlorite for 5 min.

4.   Rinse 3 × in sterile distilled water.

5.   In a laminar flow cabinet, dissect out the ovules from the pods and place them in a drop of ovule suspension medium in a Petri dish on a dissecting microscope.

6.   Isolate the embryos from the ovules using mounted needles and draw them gently in the ovule suspension medium into the tip of a Pasteur pipette.

7.   Eject the embryo, with a drop of medium, onto the surface of the G1 medium in a Petri dish (up to eight embryos per dish).

8. Cover and incubate for 10 days in darkness at 26°C, then 20 days in 12 h light : 12 h dark at 26°C.

9. In sterile conditions, transfer the embryos to R1 medium in culture tubes.

10. Incubate 12 h light : 12 h dark at 26°C until seedling development is evident and a root system has developed. At this point, seedlings may be transferred to sterile pots containing 1:1:1 sand:vermiculite:compost and grown to maturity.

Chapter 11

Regeneration of plants and micropropagation

1. Introduction

It is possible to regenerate whole plants from protoplasts, single cells and small pieces of plant tissue because plant cells are totipotent (Chapter 2). Totipotency means that plant cells can be induced through appropriate culture conditions to develop along a 'programmed' pathway leading to the formation of an entire new plant. The regenerated plant is identical to the donor plant from which the cells were derived. Plant regeneration by tissue culture is now an essential and fundamental procedure for biotechnology and plant breeding, and is used commercially for the asexual propagation of many horticultural and agricultural plants.

One of the remarkable expressions of totipotency is the unique capacity of plants to produce normal embryos from single cells in somatic tissues or callus in culture. The process is known as somatic embryogenesis and the embryos formed can be cultured *in vitro* to regenerate an entire new plant. Somatic embryogenesis was first described in carrot callus cells more than 40 years ago (Steward *et al.*, 1958; see Chapter 1). Since then protocols and procedures for somatic embryogenesis have been developed for several species and these are used both for the regeneration of plants and as a research model to study early morphogenetic and regulatory events in plant embryogenesis.

In addition to somatic embryogenesis, regeneration can be accomplished through other developmental pathways. In the simplest case, small pieces of tissue (known as explants) are excised from mature plants and under sterile conditions transferred to an appropriate nutrient culture medium. Through manipulation of the concentrations of added phyto-hormones in the culture medium, formation of a desired organ is induced (organogenesis). Typically, shoot development (caulogenesis) is induced on media formulated for the purpose. The shoots that differentiate are then transferred to a rooting medium for root development (rhizogenesis) leading to the formation of plantlets and entire plants.

2. Regeneration via somatic embryogenesis

Plants commonly reproduce through the development of zygotic embryos. Zygotic embryos develop from zygotes that are formed as a result of the events of fertilization within the embryo sac of the ovule. Through an orderly progression of cell divisions and differentiation, the embryo is formed and matures. The transition from zygote to embryo is known as zygotic embryogenesis and is the starting point of the life cycle of the plant. Further growth and development of the mature embryo is suspended as it becomes desiccated in the mature seed, but growth resumes during germination to form the emerging seedling. During embryogenesis, the pattern elements of the plant body are established including the shoot apical meristem and the root apical meristem. Almost the entire plant body is produced post-embryonically from these apical meristems (Chapter 2).

In plants, embryogenesis is not strictly dependent on fertilization. In many plant species, embryos can originate asexually either naturally through apomixis in the seed or they can be induced in tissue culture. Apomitic embryos may be derived from unfertilized eggs or from the somatic tissue of the ovule and are genetically identical to the maternal parent.

Asexual embryos can also be induced to form *in vitro* from a single cell, or a group of cells originating from a wide range of gametophytic and somatic tissues. Androgenic embryos develop from *in vitro* cultured microspores (immature pollen grains) (see Chapter 9), and somatic embryos are induced from somatic tissue in culture. In both cases, the manipulation of the concentration of plant hormones and growth regulators in the culture medium and/or the application of a stress treatment is required for embryo induction. The nature of the donor tissue and the induction treatment together determines whether embryos develop directly from single cells or indirectly via callus tissue.

The process of development of embryos from somatic cells or tissues is called somatic embryogenesis (*Figure 11.1*). The development of somatic embryos closely resembles that of zygotic embryos. All the plants derived by somatic embryogenesis from a particular tissue culture are genetically alike. Consequently, micropropagation through somatic embryogenesis is an efficient method of producing large numbers of identical elite or transgenic plants.

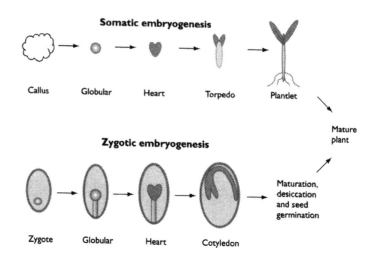

Figure 11.1

A comparison of somatic and zygotic embryogenesis showing the major stages (globular, heart, torpedo) involved.

Somatic embryos are induced from cultured callus cells by manipulation of the culture conditions. In carrot, this is a simple procedure as outlined in *Figure 11.2* and consists of the following:

(i) the establishment of a callus cell line from explants of hypocotyl tissue excised from individual seedlings, grown under sterile conditions;
(ii) transfer of cells to low auxin medium;
(iii) dilution of cells to a low density.

Not all cells in a culture are capable of forming embryos. Embryogenic competence resides within a subpopulation of the culture called pro-embryogenic cell masses (PEMs). The cells of PEMs are small with a dense cytoplasm and are full of starch grains. In contrast, non-embryogenic callus cells are large and highly vacuolated. These characteristics can be used to enrich the population of embryogenic cells in a culture through selecting the PEMs out from the culture by sieving or density gradient fractionation.

Somatic embryos pass through the same sequence of characteristic morphological stages as zygotic but lack a desiccation phase and grow directly into plantlets (*Figure 11.2*). The first recognizable stage is the globular stage, in which the embryo is spherical. In zygotic embryogenesis, the globular embryo is attached to the maternal tissue by the suspensor. The globular stage of somatic embryos grows from PEMs within 5–7 days after the cultures have been transferred to a low auxin medium. In contrast to embryos developing in liquid culture, embryos grown on a solid medium may develop a small suspensor-like region. After 2–3 days of isodiametric growth, the globular stage changes to bilaterally symmetrical growth to form an oblong structure. This is the beginning of the heart stage, which is characterized by the

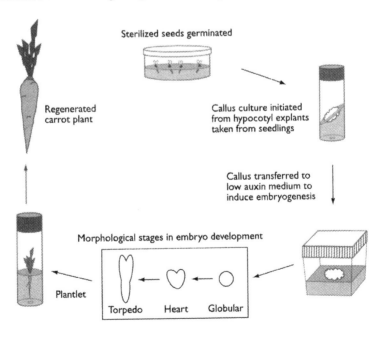

Figure 11.2

Stages of induction of viable embryos via callus in carrot.

expansion of the two cotyledons. The heart stage is followed by the torpedo stage clearly marked by the elongation of the hypocotyl and the beginning of radicle formation. Approximately 18–21 days after induction, plantlets are discernible. These plantlets have green cotyledons, elongated hypocotyls and radicles with clearly developed root hairs. These plantlets can then be transferred to solid media to complete the regeneration of whole plants.

3. Control of embryogenesis

The embryogenic competence of a culture is at its highest when the culture is relatively young, i.e. up to a year after its initiation. However, cultures that have been repeatedly subcultured and maintained in an undifferentiated state at high auxin concentrations for many years may progressively lose their embryogenic capacity.

Auxin is the most important growth regulator involved in both the induction of somatic embryogenesis and the proper morphogenic development of the embryo. Carrot cell cultures require the auxin analogue 2,4-D for embryogenic competence, but the continuous presence of this auxin blocks further development beyond the PEM stage and in some cell lines up to but not beyond the globular stage. It follows that removal of this block by shifting the culture to low auxin or to auxin-free media induces embryogenesis. As embryogenesis proceeds, the developing embryos begin to synthesize their own auxins. Furthermore, the polar transport of auxin is necessary for normal morphogenesis beyond the globular stage. In addition to added auxins, other exogenous molecules are implicated in somatic embryogenesis. These include proteins that are secreted into the culture medium such as endochitinases and arabinogalactan (AGPs) proteins.

Details of the molecular events that occur during the induction and progress of somatic embryogenesis are largely lacking. However, many attempts are being made to identify specific genes that control early events in embryogenesis. These genes would be useful as research tools to study in detail the transition of somatic cells into embryogenesis. Furthermore, these genes and/or their products could find application as specific markers suitable for the manual or automated separation of embryogenic from non-embryogenic cells in mixed cell populations. So far, the majority of genes identified appear to control basic developmental processes and are not restricted to the embryo. However, more recently, a gene has been identified that is expressed in small cell populations of carrot cultures during the initiation of embryogenesis. This gene, a somatic embryogenesis kinase (DcSERK: GenBank accession no. A67796) encodes a Leu-rich repeat (LRR) transmembrane receptor-like kinase (RLK). DcSERK is expressed in early somatic embryos and in globular zygotic embryos. DcSERK and its *Arabidopsis* analogue AtSERK1 are thought to be part of a signalling pathway that switch on an embryogenesis development programme.

4. Regeneration via organogenesis

For many species, it is possible to regenerate plants from meristematic tissues. For example, the formation of multiple shoots can be induced from shoot tip and axillary bud explants by using tissue culture media with defined phytohormone concentrations and/or combinations. The many shoots that are formed can be separated and grown on a suitable rooting medium to produce plantlets.

In some species, it is also possible to regenerate plants directly by organogenesis from explants of adventive and non-meristematic tissue or indirectly through callus tissue. As a general principle, the regeneration is induced by variations in auxin/cytokinin ratio.

The regeneration of plants from tissue explants by organogenesis is extensively used commercially for the micropropagation of economically important plants. Typically, this procedure consists of four stages:

(i) Initiation stage. A selected explant (a piece of tissue) is excised from a selected donor plant, surface sterilized and transferred to a solid culture medium to establish an aseptic culture.

(ii) Multiplication stage. A growing explant is induced to produce vegetative shoots by transfer to a medium containing cytokinin

(iii) Rooting or pre-plant stage. The differentiated shoots are induced to produce adventitious roots by shifting to a medium that contains an auxin. For easily rooted plants, an auxin is usually not necessary and many commercial protocols omit this step.

(iv) Acclimatization. Growing, rooted shoots are removed from tissue culture medium and placed in soil. These plantlets are developed under carefully controlled conditions where they are gradually acclimatized to conditions such as low humidity as tissue-cultured plants are extremely susceptible to wilting.

5. Protocols

A general protocol is presented for the regeneration of viable seedlings from callus of both monocots and dicots, using transfer from a high-auxin, callus-inducing medium to a hormone-free medium. In our laboratory, we grow callus in Petri dishes, then initially transfer embryos to culture tubes (Universal bottles or plastic equivalents) before moving growing seedlings on into larger vessels (Kilner jars or plasticware; see *Figures 11.3* and *11.4* and Chapter 5 for details). A protocol for *Arabidopsis* regeneration based on the method of Mather and Koncz (1998) is also presented. This is a commonly used method for transformed plants in which shoots are initiated first, followed by roots. In our hands, we expect to obtain viable plants about 3 months after callus initiation using this type of method. In addition, we briefly summarize a method for clonal propogation of gymnosperm trees (Attree *et al.*, 1996). Techniques based on this have been successfully commercialized for the bulk production of clonal trees (Chapter 3). An example of a gymnosperm embryo grown from callus is shown in *Figure 11.4*).

(a)

(b)

Figure 11.3

A Nicotiana *plantlet grown in sterile conditions on MS agar in sterile conditions in a glass jar 14 days after transfer to the jar (a) and after 6 weeks (b). The plantlet in (b) is ready to be acclimated to a glasshouse.*

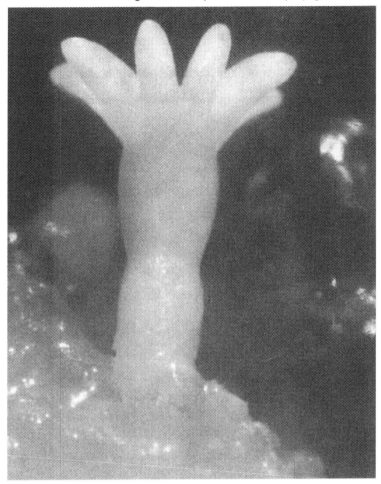

Figure 11.4

A Norway spruce embryo produced directly from embryogenic callus.

Further reading

Greyson, R.I. and Walden, D.B. (1994) Axillary bud *in vitro* culture: asexual propagation of maize. In: *The Maize Handbook* (eds. M. Freeling and V. Walbot). Springer Verlag, New York.

Zimmerman, J.L. (1993) Somatic embryogenesis: a model for early development in higher plants. *Plant Cell* **5**: 1411–1423.

References

Attree, S., Rennie, P.J. and Fowke, L.C. (1996) Induction of somatic embryogenesis in conifers. In: *Plant Tissue Culture and Laboratory Exercises* (eds. R.J. Trigiano and R.J. Gray). CRC Press, Boca Raton, Florida.

Mather, J. and Koncz, C. (1998) Callus culture and regeneration. In: *Methods in Molecular Biology*, Vol. 82, Arabidopsis *protocols* (eds. J. Martinez-Zapater and J. Salinas). Humana Press, Totowa, New Jersey.

Steward, F.C., Mapes, M.O. and Mears, K. (1958) Growth and organized development of cultured cells. *Am. J. Bot.* **45**: 705–708.

Protocol 11.1

Embryogenesis from callus in a dicot (e.g. carrot) or monocot (cereals, rice)

Equipment

Binocular microscope

Growth cabinet (25°C, 16 h light : 8 h dark).

Materials and reagents (sterile)

Embryogenic cultures growing on (Dicot) MS medium plus 2–5 mg l^{-1} 2,4-D, 0.6 mg l^{-1} kinetin and 2% sucrose (Monocot) MS medium plus 0.5 mg l^{-1} 2,4-D and 3% sucrose (3–6 weeks old, then subcultured every 4 weeks). See *Protocols 6.2* and *6.3*

Ethanol 95%

Liquid MS medium 150 ml

Petri dishes containing MS medium minus hormones + 0.8% agar + 2% sucrose

Universal tubes or similar (see Chapter 5) containing slopes of MS agar without hormones + 2% sucrose + 0.8% agar

Kilner jars or plasticware (see Chapter 5) containing MS agar without 2,4-D or cytokinin + 2% sucrose

Plasticware or pots with translucent covers containing sterile potting compost

Sterile dissecting instruments.

Protocol

1. Remove 0.5–1.0 cm^2 pieces of callus to the surface of the MS agar without hormones.

2. Incubate in 16 h light : 8 h dark at 25°C.

3. Embryos visible first as green areas will begin to form after 2 weeks on the surface of the callus. Under the dissecting microscope, globular, heart-shaped and torpedo-shaped embryos will be seen.

4. Carefully remove the embryos and subculture them onto MS agar in the absence of hormones in Universal tubes until a viable seedling is formed.

5. Subculture seedlings to Kilner jars or equivalent plasticware until four to five leaves are formed.

6. Wash the roots in fungicide (e.g. Benomyl, 0.02%) and transfer to sterile potting compost in a humid container. Gradually reduce humidity until the seedling is fully acclimated.

Protocol 11.2

Plant regeneration by organogenesis

Equipment

Binocular microscope

Orbital incubator at 120 rpm, 25°C, 16 h light : 8 h dark

Incubator or growth room 25°C, 16 h light : 8 h dark

Microfuge

Universal tubes

Kilner jars

Sterile forceps, scalpel, scissors

9-cm Petri dishes.

Materials and reagents

10% sodium hypochlorite with 0.1% detergent (Tween, Triton)

Sterile distilled water

0.5 MS medium, 0.8% agar (pH 5.8) (in 9 cm Petri dishes)

MS medium with 3% sucrose (pH 5.8)

0.5 MS medium with 0.5% sucrose and 0.8% agar (in Kilner jar or equivalent)

Callus medium: MS medium with 3% sucrose (pH 5.8) + 2,4-D, 0.5 mg l^{-1}; IAA, 2.0 mg l^{-1}; IPAR, 0.5 mg l^{-1}; NAA, 0.5 mg l^{-1}; 0.8% agar (in 9-cm Petri dishes)

Shoot medium: MS medium with 3% sucrose (pH 5.8) + IPAR, 2.0 mg l^{-1}; NAA, 0.05 mg l^{-1}; 0.8% agar (in 9 cm Petri dishes and Universal tubes or Kilner jars or similar plasticware)

Root medium: MS medium with 3% sucrose (pH 5.8) + IAA, 1.0 mg l^{-1}; IBA, 0.2 mg l^{-1}; IPAR, 0.2 mg l^{-1}; 0.8% agar (in Universal tubes or Kilner jars or similar plasticware).

Protocol

1. Surface-sterilize 0.1 g seeds in sodium hypochlorite in an Eppendorf tube for 15 min with shaking.

2. Pellet by brief centrifugation, pour off the supernatant and wash 5 × with sterile distilled water.

3. Germinate the seeds in Petri dishes containing 0.5 × MS medium 0.8% agar in 16 h light : 8 h dark at 25°C.

4. After 1 week, place seedlings in 250-ml flask with 35 ml of MS medium with 3% sucrose and no agar. Cap and place on orbital shaker at 120 rpm, 16 h light : 8 h dark 25°C for 15–20 days (this gives a proliferation of root material).

5. Harvest the roots by cutting them from the seedlings.

6. Place on callus medium in a Petri dish and incubate at 25°C.

7. When callus forms (2–3 weeks) transfer to shoot medium in a Petri dish and incubate for 2 weeks until shoots form.

8. Transfer to shoot medium in a Universal tube or similar.

9. When 4–6 leaves have formed, transfer to root medium in a Universal tube or Kilner jar or plastic equivalent until roots form.

10. Finally, allow to form small seedlings in a larger Kilner jar or equivalent on 0.5 MS medium 0.8% agar and 0.5% sucrose.

Protocol 11.3

Somatic embryogenesis of Norway spruce using a suspension culture step

Equipment

As *Protocol 6.5*

Orbital incubator (150 rpm 20–25°C)

Conical flasks (250-ml)

Aluminium foil caps (sterile)

Universal tubes or similar plasticware

Fine nylon mesh sterile

9-cm filter paper (sterile)

9-cm Petri dishes

Desiccator containing saturated ammonium nitrate (yields a relative humidity of 63%).

Materials and reagents

As *Protocol 6.5* and:

Maintenance medium: 0.5 strength Litvay's medium with L-glutamine, 250 mg l^{-1}; casein hydrolysate, 500 mg l^{-1}; 2,4-D, 2 mg l^{-1}; benzyladenine 1 mg l^{-1}; sucrose 1%

Wash medium: 0.5 strength Litvay's medium with L-glutamine, 250 mg l^{-1}; casein hydrolysate, 500 mg l^{-1}; sucrose 3%

Development medium: 0.5 strength Litvay's medium with L-glutamine, 250 mg l^{-1}; casein hydrolysate, 500 mg l^{-1}; sucrose 3%, PEG 4000, 7.5%; ABA 4 µm; agar 0.8%

Plantlet conversion medium: 0.5 strength Litvay's medium with L-glutamine, 250 mg l^{-1}; casein hydrolysate, 500 mg l^{-1}; sucrose 2%; agar 0.6%

Gymnosperm callus prepared as in *Protocol 6.5*.

Protocol

1. Select white, translucent embryogenic tissue from the callus obtained in *Protocol 6.2*.

2. Subculture to fresh plates as in *Protocol 6.2* every 2–4 weeks for two to four subcultures.

3. Transfer rapidly growing embryogenic callus to a 250-ml conical flask containing 20–50 ml of maintenance medium. Incubate on orbital incubator at 150 rpm.

4. Replace medium every week by allowing cells to settle and withdrawing spent medium.

5. After 4–6 weeks, remove an aliquot of suspension to 50 ml of fresh maintenance medium.

6. Repeat, reducing the amount of inoculum weekly (usually to 10 ml in 50 ml of medium) until an embryogenic suspension appears (test different volumes of inoculum in several flasks and choose the one giving most embryogenic activity).

7. Sieve the cell suspension in a fine nylon mesh. Transfer 10 g sieved cells to 50 ml of wash medium in a 250-ml flask.

8. Swirl constantly and pipette onto filter paper supports floating on the surface of development medium in Petri dishes.

9. Transfer the filter paper to fresh development medium every 2–4 weeks. Maintain for 6–8 weeks. By this stage, embryos at the cotyledonary stage will have developed.

10. Transfer filter paper supports with embryos to empty sterile Petri dishes in a desiccator at 63% relative humidity (see *Equipment*) for 2 weeks.

11. After desiccation, embryos may be stored at −20°C or transferred to plant-let conversion medium while still on the filter paper support. They should be maintained under low light for 1–2 days.

12. Select individual embryos and place on plantlet conversion medium in Universal tubes or similar for 4–6 weeks.

13. Wash roots free of medium and place in sterile soil in pots or trays with translucent covers. Maintain a high relative humidity for the first week.

Chapter 12

Somaclonal variation

1. Introduction

The culture regimes used for the regeneration of whole plants from tissue explants, undifferentiated cells and protoplasts are frequently a source of variability. This means that a significant percentage of the regenerated plants may not be identical in genotype and phenotype to the plant from which the original explant was obtained. Larkin and Scowcroft (1981) named this phenomenon somaclonal variation.

The origin and expression of the observed variation differs from species to species as well as the specific sources of the tissue explants.

2. Origins and mechanisms of somaclonal variability

Somaclonal variation can be of two sorts:

(i) genetic (i.e. heritable) variability – caused by mutations or other changes in DNA;
(ii) epigenetic (i.e. non-heritable) variability – caused by temporary phenotypic changes.

2.1 Genetic variability

Various molecular mechanisms are responsible for genetic variability associated with somaclonal variation.

Changes in ploidy

One of the more frequently encountered types of somaclonal variation results from changes in chromosome number, that is, aneuploidy, polyploidy or mixoploidy. Changes in ploidy originate from abnormalities that occur during mitosis. For example, extra chromosomal duplication during interphase, spindle fusion or lack of spindle formation and cytoplasmic division.

As plant cells grow and age, the frequency of changes in ploidy increases. Therefore, changes in ploidy observed in cultures and regenerated plants might have their origins in the source of tissue explants used.

Another cause of variability due to changes in ploidy is the *in vitro* culture regime itself. The longer the cells remain in culture the greater is their chromosomal instability. In addition, the composition of the growth medium can trigger changes in ploidy. For example, both kinetin and 2,4-D are implicated in ploidy changes and cultures grown under nutrient limitations can develop abnormalities. Selecting a suitable explant and an appropriate culture medium can therefore enhance the chromosomal stability of the culture. However, high variations of ploidy in cultures do not always lead to high frequencies of somaclonal variation in regenerated plants. This is because, in mixed cultures, diploid cells appear to be better fitted than aneuploid or polyploid cells for regeneration, as they are more likely to form meristems.

Structural changes in nuclear DNA

Structural changes in nuclear DNA appear to be a major cause of somaclonal variation. The changes can modify large regions of a chromosome and so may affect one or several genes at a time. These modifications include the following gross structural rearrangements (*Figure 12.1*):

- deletions – loss of genes;
- inversions – alterations in gene order;
- duplications – duplication of genes;
- translocations – segments of chromosomes moving to new locations.

Activation of transposons can be a cause of somaclonal variation. Transposons or transposable elements are mobile segments of DNA that can insert into coding regions and cause gene disruption.

In addition to these larger modifications of nuclear DNA sequence, changes at the level of a single DNA nucleotide that occur in a coding region can lead to somaclonal variation. For example, point mutations that result from a change of base in a single nucleotide or the altered methylation of a base can lead to gene inactivation.

Chimaeral rearrangement of tissue layers

Many horticultural plants are periclinal chimaeras, that is, the genetic composition of each concentric cell layer (LI, LII, LIII) of a meristem (e.g. the shoot tip meristem; *Figure 12.2* and Chapter 2) is different. These layers can be rearranged during rapid cellular proliferation. Therefore, regenerated plants may contain a different chimaeral composition or may no longer be chimaeric at all. Shoot tip transformation procedures are particularly likely to cause chimaeral transgenics.

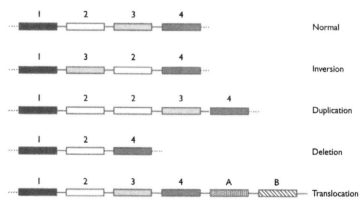

Figure 12.1

Patterns of gross DNA changes that can cause somaclonal variation. Product 1 is generated by the arbitrary primers and product 2 by primers B and D. Products from the PCR reaction are separated by electrophoresis and the pattern further analysed.

Figure 12.2

The organization of layers L1–L3 in a shoot apical meristem. The meristem is viewed in transverse section; layer L1 forms the epidermis, layer L2 the subepidermal layer and layer L3 the internal tissues of the shoot.

2.2 Epigenetic variability

Epigenetic changes that cause somaclonal variation can be temporary and over time are reversible. However, sometimes they can persist through the life of the regenerated plant. One common phenotypic change seen in plants produced through tissue culture is rejuvenation. Rejuvenation causes changes in morphology such as earlier flowering or enhanced adventitious root formation. Epigenetic changes may be caused by DNA methylation, DNA amplification or by activation of transposable elements (transposons). DNA methylation is important for both transcription and translation because it alters the substrate affinities of enzymes that are active in these processes. Tissue culture media can change the level of DNA methylation and thus may be one of the important causes of somaclonal variation.

3. The importance of somaclonal variation

The *in vitro* procedures for transformation, regeneration and clonal propagation of plants may involve tissue culture procedures that can take anywhere from 4 weeks to several months to complete. The degree of somaclonal variation generated and accumulated during this period of culture can be significant and is an unwelcome interference that is difficult to control. In plant transformation procedures, apart from the mutagenic effects of the basic *in vitro* culturing process, methods of DNA delivery and selection may also induce additional somaclonal variation.

Extensive somaclonal variation can result in undesirable changes in the genetic background of regenerated plants that could limit their use as cultivars or as elite parents in breeding programmes. It is therefore important to assess the genetic variability induced during transformation and/or regeneration of plants through tissue culture.

4. Methods of assessing somaclonal variation in regenerated plants

Although cytological and phenotypic analyses can be used to evaluate somaclonal variation, recently molecular techniques have been used with increasing frequency.

4.1   Restriction fragment length polymorphism (RFLP)

RFLP was one of the first techniques to be applied to somaclonal variation and has been widely used for several species. RFLP is a hybridization-based technique that detects variation at the DNA sequence level but requires the use of probes that hybridize to known sequences.

A number of amplification techniques based on PCR technology have now been developed that avoid the need for prior sequence information.

4.2   Random amplified polymorphic DNA-PCR (RAPD-PCR)

RAPD-PCR or arbitrarily primed PCR (AP-PCR) (Williams *et al.*, 1990) is a technique that has proved useful in detecting somaclonal variation in a number of species (e.g. Saker *et al.*, 2000). RAPD-PCR is based on the premise that, because of its complexity, eukaryotic nuclear DNA may contain paired random segments that are complementary to single decanucleotides and furthermore these segments have the correct orientation and are located close enough to each other for PCR amplification (*Figure 12.3*). RAPD-PCR uses single primers of arbitrary nucleotide sequence to initiate DNA synthesis. The DNA fragments can be separated by gel electrophoresis and the DNA variation is detected by the pattern of DNA bands from individual plants.

4.3   Amplified fragment length polymorphism (AFLP)

More recently, another PCR-based technique known as AFLP (Vos *et al.*, 1995) has been used to study somaclonal variation (Polanco and Ruiz, 2002). AFLP is a DNA-fingerprinting procedure based on the selective PCR amplification of fragments from restriction digestion of genomic DNA.

AFLP analysis consists of the following steps (*Figure 12.4*).

- Genomic DNA is digested with two restriction enzymes, one that cuts frequently, for example *Mse*I (4-bp recognition sequence) and one that cuts less frequently, for example *Eco*RI (6-bp recognition sequence).
- The resulting fragments are ligated to double-stranded adapter molecules that consist of a core sequence and a sequence specific for either the *Eco*RI site or the *Mse*I site.
- Pre-selective amplification by PCR using primers designed to include a core sequence, an enzyme specific sequence and a single base extension

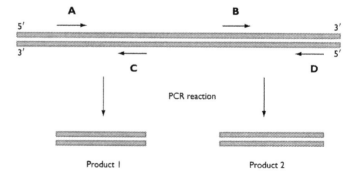

Figure 12.3

Illustration of the principle of RAPD.

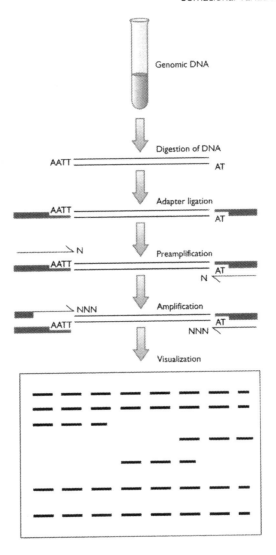

Figure 12.4

Illustration of the principle of AFLP.

at the 3′ end. The primary products are fragments containing one *Mse*I cut, one *Eco*RI cut and a matching internal nucleotide. This amplification step achieves a 16-fold reduction in complexity.

■ Selective PCR amplification using the products of the pre-selective amplification as template and identical primers as the pre-selective step, but containing two further additional nucleotides at the 3′ end. The primers are either radio-labelled or fluorescently labelled. Only fragments having the matching nucleotides in all three positions (50–200) will be amplified. This step reduces the complexity 250-fold.

■ Gel electrophoretic separation reveals the pattern (fingerprint) of the labelled fragments that are analysed with the aid of an appropriate software package.

The molecular analyses described above are rapid and more precise than phenotypic analyses. However, both the RAPD and AFLP techniques have proved to be inconclusive in some studies. Detailed protocols for AFLP are available at http://www.dpw.wau.nl/pv/index.htm.

5.   Somaclonal variation as a technique for crop improvement

The recognition that cell culture regimes were sources of somaclonal variation in regenerated plants led to the use of cell culture for mutagenesis and the direct selection of genetic variants with valuable traits. Such a selection strategy was considered a particularly effective way of obtaining variants that were tolerant to environmental factors such as cold, drought, salinity or toxic ions or resistance to pathogens. Although there are cases where somaclonal variation has produced plants with useful agricultural traits, as a source of variation somaclonal variation is difficult to exploit because the variation is largely unstable, mostly unpredictable and can be deleterious. This technique has not proved useful as a source of genetic variation for plant improvement as was once anticipated.

Further reading

Skirvin, R.M., Coyner, M., Norton, M.A., Motoike, S. and Gorvin, D. (2000) *AgBiotechNet*, Vol. 2, ABN 048.

References

Larkin, P.J. and Scowcroft, W.R. (1981) Somaclonal variation – a novel source of variability from cell cultures. *Theor. App. Genet.* **67**: 197–201.

Polanco, C. and Ruiz, M.L. (2002) AFLP analysis of somaclonal variation in *Arabidopsis thaliana* regenerated plants. *Plant Sci.* **162**: 817–824.

Saker, M.M., Bekheet, S.A., Taha, H.S., Fahmy, A.S. and Moursy, H.A. (2000) Detection of somaclonal variations in tissue culture-derived date palm plants using isoenzyme analysis and RAPD fingerprints. *Biol. Plant.* **43**: 347–351.

Vos, P., Hogers, R., Blecker, M., Van de Lee, T., Hornes, M., Frijters, A., Pot, J., Peleman, J., Kuiper, M. and Zabeau, M. (1995) AFLP: A new technique for DNA fingerprinting. *Nucleic Acids Res.* **23**: 4407–4414.

Williams, J.G.K., Kubelik, A.R., Livak, K.J., Rafalski, J.A. and Tingey, S.V. (1990) DNA polymorphisms amplified by arbitrary primers as useful genetic markers. *Nucleic Acids Res.* **18**: 6231–6235.

Chapter 13

Bacterial culture in the plant cell culture laboratory

1. Introduction

The production of transgenic plants requires methods for identifying, isolating, replicating and inserting DNA into the plant genome. All these steps may involve the use of bacteria and bacterial culture (especially of *E. coli* and *Agrobacterium tumefaciens*) is routine in the plant tissue culture laboratory undertaking plant transformation. Thus while the culture techniques required are different from those for plant and cell culture, a basic description will be given here.

A vector (which may be either bacterial or viral) is needed to introduce a gene into a foreign genome. The vector system most commonly used for plant transformations is based on the tumour-inducing plasmid of *A. tumefaciens*. Types of transformation vectors are described in detail in Chapter 4. Prior to insertion into a transformation vector, the isolated gene must be integrated into a cloning vector – a bacterial plasmid which can be multiplied in a bacterium like *E. coli* that is easy to handle in the laboratory. *E. coli* and *A. tumefaciens* are available as modified strains that minimize risk to the environment. *E. coli* cells treated with calcium chloride will take up plasmids from the culture medium and replicate them. They can be grown to logarithmic phase, incubated in calcium chloride and snap-frozen as competent cells – ready to take up a plasmid when required (*Protocol 13.1*). Fragments of DNA can also be packaged into bacteriophage λ, which can be used to transform the bacteria.

Agrobacterium rhizogenes is also used as a transformation system for plant cell cultures, where it produces 'hairy' roots – roots with abundant root hairs. These have been used in tissue culture as they are effectively immortal and produce abundant and rapidly growing root cultures (Chapter 10). Techniques for the culture of *A. rhizogenes* are therefore also included with those for *E. coli* and *A. tumefaciens*.

2. Facilities for bacterial culture

To culture bacteria successfully, a clean bench in a room that is not a through room is recommended. A Bunsen burner can be used to sterilize tools and to flame the tops of vessels before transfer takes place. It also serves to provide a circular area on the bench that can be regarded as sterile to work in. Glass spreaders should be dipped in 70% alcohol and held in the Bunsen flame until all the alcohol has burnt off to sterilize them. When holding these it is important to hold the head of the spreader at a lower position than your hand so that burning alcohol cannot flow on to you. These must be cooled before use. To sterilize a wire loop, it should be heated until red-hot before cooling. In contrast to the spreader, the head of the loop should be held up in relation to the hand so that the handle does not heat up and burn the operator. An alternative to using loops to inoculate a liquid culture is to use an automatic pipettor and sterile tips. Small, open containers lined with an autoclave bag may be used for disposal of these tips. Both the loop and the glass spreader should be sterilized by Bunsen immediately after use.

To culture bacteria, both a stationary incubator for plates and an orbital, shaking incubator for liquid cultures, set at 28°C for *Agrobacteria* or 37°C for *E. coli*, are necessary. For bulking up bacterial cultures, orbital shakers that are capable of holding 2-l conical flasks should be available. Most modern platform shakers are flexible enough to cope with a range of sizes and shapes of vessels by supplying racks and clamps. Bacterial cultures do not require illuminated growth facilities. To deal with solid, contaminated waste, a metal bucket suitable for the autoclave should be under the bench. Plastic boxes, also suitable for the autoclave are available. Both containers should be lined with an autoclave bag before use. For liquid-contaminated waste, a jug of hypochlorite solution or disinfectant like Virkon™ should be on the bench or on a low platform very close by. These should be renewed daily. Powdered Virkon™ and a source of paper towel should be close at hand in case of a liquid spill. It is essential to have access to an autoclave in order to sterilize your media and equipment, and also to dispose of contaminated waste, but it should not be inside the culture facility.

Other general equipment for bacterial culture includes a pH meter, source of ultrapure water, magnetic stirrer, balance, vortex mixer and cuvettes for spectrophotometry (glass or disposable plastic for use at visible wavelengths and quartz for UV spectra). Gloves and water- and alcohol-proof marker pens are also essential. A microwave oven can be used to melt agar stocks and a water bath set at 45°C can be used to maintain molten agar until it is poured. However, never place sealed containers in a microwave oven; even containers with a lightly screwed lid can seal and generate explosive pressure.

Further reading

Jones, P.G. and Sutton, J.M. (1997) *Plant Molecular Biology: Essential Techniques*. John Wiley, Chichester.

Protocol 13.1

o culture *E. coli* in Luria broth (LB)

Equipment

Bacterial wire loop

Bunsen burner

Shaking incubator set at 37°C

Spectrophotometer set at 600 nm

Timer

Sterile glass universal bottles and tops

Sterile 250-ml Erlenmeyer flask sealed with cotton wool.

Materials and reagents

70% alcohol to clean the bench

Sterile Luria broth: tryptone, $10 \, g \, l^{-1}$; yeast extract, $5 \, g \, l^{-1}$; NaCl, $10 \, g \, l^{-1}$; D-glucose, $10 \, g \, l^{-1}$, pH 7.0

Plates of LB agar (1%) streaked with the *E. coli* strain of choice 24 h prior to use.

Protocol

1.   Sterilize wire loop by heating in Bunsen flame until it is red hot. Allow to cool.

2.   Inoculate a single colony by loop from a fresh plate of the *E. coli* strain into 10 ml of LB and grow overnight at 37°C in a shaking incubator at 150 rpm.

3.   In the morning, inoculate 500 µl of overnight culture into 50 ml of fresh LB and grow for 90 min to an optical density of 0.1 at 600 nm. These should be in the early phase of logarithmic growth and will be ready for transformation or for further use.

Protocol 13.2

Calcium chloride-mediated transformation of *E. coli*

Equipment

      Bench centrifuge and centrifuge tubes

      Glass spreader

      Glass universal bottles and 200-ml flasks

      Ice in ice bucket

      Timer

      Micropipettor and tips

      Pipettes

      Shaking incubator set at 37°C

      Spectrophotometer

      Water bath at 42°C.

Materials and reagents

      Antibiotic for selection of *E. coli* transformed with plasmid

      100 mM $CaCl_2$

      50 mM $CaCl_2$

      E. coli strain (suitable for high-frequency transformation, e.g. DH5α)

      Ethanol

      LB

      LB supplemented with 1% agar

      Plasmid DNA

      Tris-EDTA buffer (TE): 10 mM Tris-HCl, pH 7.5, 1 mM EDTA.

Protocol

1. Inoculate a single colony from a fresh plate of the *E. coli* strain into 10 ml of LB and grow overnight at 37°C in a shaking incubator.

2. Inoculate 500 µl of overnight culture into 50 ml of fresh LB and grow for 90 min, to an absorption at 600 nm of approximately 0.1.

3. Centrifuge the cells in a cooled centrifuge for 10 min at 4000 *g*.

4. Pour off the supernatant and gently resuspend the pellet in 15 ml of ice-cold 100 mM $CaCl_2$. Incubate the cells on ice for 30 min. *E. coli* cells rapidly lose

their competence to take up DNA if they are kept warmer than 4°C, or if they are not treated gently.

5. Centrifuge the cells at full speed in a cooled bench centrifuge, pour off the supernatant and gently resuspend the pellet in 2 ml of ice-cold 50 mM $CaCl_2$. (At this stage of the procedure, the cells are described as competent. They may either be snap-frozen in liquid nitrogen and stored for up to 3 months or they may be left on ice for a minimum of 2 h then transformed as in the next steps.)

6. Add up to 100 µl of plasmid DNA to 200 µl of competent cells and incubate on ice for 30 min.

7. Heat-shock the cells by placing the tubes in a 42°C water bath for 2 min. Return the tubes to the ice bucket and allow to recover for 5 min.

8. Transfer the cells to a glass universal bottle. Add 1 ml of LB and grow for 1 h at 37°C in a shaking incubator.

9. Prepare LB agar plates with the appropriate antibiotic to select for the transformed *E. coli* cells.

10. Plate 50 µl of transformed cells on to the selective plate and spread smoothly, using a flamed glass spreader.

11. Incubate the plates at 37°C overnight or until the colonies are visible.

12. Pick single, well-isolated colonies and check for the presence of the plasmid using standard molecular biology techniques.

Protocol 13.3

Calcium chloride-mediated transformation of *Agrobacterium*

Equipment

Bench centrifuge and centrifuge tubes

−70°C freezer

Ice in an ice bucket

Automatic pipettor and sterile tips

Shaker set at 28°C

Spectrophotometer

Water bath set at 37°C

Sterile freezing tubes.

Materials and reagents

Agrobacterium strain of choice

Antibiotics for selection of binary construct

20% v/v glycerol

Liquid nitrogen

TE medium: 10 mM Tris-HCl, pH 8.0, 1 mM EDTA

TY medium: 0.5% tryptone, 0.3% yeast extract, 0.13% $CaCl_2$, pH 7.0

TY agar: TY medium supplemented with 1% agar

Vector DNA from binary construct.

Protocol

1. Inoculate a 10-ml culture of TY medium from a single colony of *Agrobacterium* and grow in a shaking incubator at 28°C overnight.

2. Inoculate 200 μl of the culture into 200 ml of TY medium and shake at 28°C until the culture reaches an absorbance of 0.4 at 600 nm (8–12 h).

3. Centrifuge the culture in a cooled centrifuge at 4000 **g** for 10 min at 4°C. Pour off supernatant and wash pellet with 20 ml of TE.

4. Repeat centrifugation step and resuspend the pellet in 20 ml of ice-cold TY medium.

5. Dispense the cells into 500-μl aliquots in 1.5-ml freezing tubes and immediately freeze in liquid nitrogen. Care must be taken in using liquid nitrogen. At this stage, the competent cells may be either frozen in a −70°C freezer and

stored for up to 3 months, or thawed on ice and transformed as in the subsequent procedure.

6. After thawing, add approximately I µg of binary construct DNA in a volume not exceeding 50 µl and mix gently.

7. Incubate the cells for 5 min on ice.

8. Transfer the tubes to liquid nitrogen for 5 min and then plunge them into a 37°C water bath for 5 min.

9. Add I ml of TY medium and shake the cultures at 28°C for 4 h.

10. Plate the transformed cells on TY plates containing antibiotics for the selection of the binary construct. Grow at 28°C for 24–48 h.

11. Pick single colonies and restreak on to new plates.

12. Store the resulting strain by resuspending bacteria from the plate in sterile 20% glycerol and freezing at –70°C.

Protocol 13.4

Culture of *Agrobacterium tumefaciens* in culture medium, YEB

Equipment

Incubator set at 28°C

Shaking incubator set at 28°C

Bacterial wire loop

Spectrophotometer set to 600 nm

Bench centrifuge and centrifuge tubes.

Materials and reagents (all sterile)

YEB: 10-ml aliquots sterilized in glass universal bottles; yeast extract, 1 g l^{-1}; beef extract, 5 g l^{-1}; peptone, 5 g l^{-1}; sucrose, 5 g l^{-1}; MgSO$_4$.7H$_2$O, 0.5 g l^{-1}; pH 7.0

YEB: 50 ml sterilized in 250-ml Erlenmeyer flask, sealed with cotton wool

YEB agar plates

Antibiotics.

Protocol

1.   Inoculate a single colony of the A. *tumefaciens* strain from a fresh plate into a 10-ml culture of YMA liquid medium supplemented with antibiotics appropriate to the strain and construct resistance.

2.   Grow overnight at 28°C in a rotary shaker at 130 rpm.

3.   Next morning, inoculate 500 µl of the overnight culture into 50 ml of fresh YEB without antibiotics.

4.   Grow this overnight at 28°C, shaking at 130 rpm until the culture is in the late logarithmic phase. The optical density at 600 nm should be around 0.8.

5.   This culture can be centrifuged at 5000 g for 10 min, washed in appropriate plant culture medium, resuspended to an O.D. of 2.0 and used for transformation of leaf discs.

6.   A. *tumefaciens* is sensitive to the antibiotic carbenicillin and the combination antibiotic timentin (clavulanic acid/ticarcillin, Melford Laboratories).

Protocol 13.5

Culture of *Agrobacterium rhizogenes* in culture medium, YMA

Equipment

Incubator set at 28°C

Sterile syringe needle.

Materials and reagents

YMA broth plus 1% agar plates: $MgSO_4.7H_2O$, 2 g l^{-1}; K_2HPO_4, 0.5 g l^{-1}; NaCl, 0.1 g l^{-1}; mannitol, 10 g l^{-1}; yeast extract, 0.4 g l^{-1}; pH 7.0

Antibiotics

Plant tissue.

Protocol

1. Inoculate the chosen *Agrobacterium* strain on to a fresh plate of YMA plus antibiotics appropriate to the strain and incubate overnight at 28°C.

2. When using to transform plant tissue, dip the tip of a sterile syringe needle into the culture of A. *rhizogenes* and use to stab the sterile plant tissue at 2–3-mm intervals.

3. The infected plant material should be incubated under conditions suitable for the plant tissue. Hairy roots should emerge within 14 days.

4. A. *rhizogenes* is sensitive to the antibiotic Cefotaxime.

Chapter 14

Industrial uses of plant cell culture

1. Introduction

The potential of plant cell culture for the production of high-value compounds was first recognized in the 1950s and resulted in intensive research activity in the 1970s and 1980s. Recently, interest has tended to focus instead on engineering transgenic crops to produce such materials. The prospect of avoiding the high costs of industrial plant, its vulnerability to infection and the direct use of solar energy has made such crops seem attractive. However, interest in industrial production of high-value products remains, together with the use of plant cells for biotransformations. This chapter will introduce the use of plant cells on an industrial scale, together with some examples of industrial production using them.

2. Plant cells in culture

Plant cells in culture show a number of properties that cause particular difficulties in large-scale cell culture systems. *Table 14.1* compares the key features of plant and microbial cells. Plant cells are larger than microbes and, perhaps consequently, are much more susceptible to physical damage. They are also more sensitive to osmotic effects. Plant cell cultures grow slowly – perhaps 5–10 times more slowly than microbial cultures. As a consequence, they are much more susceptible to contaminant organisms. Yields will be lower, because of slower growth rates, and because frequently, the high-value product desired is produced not during the growth phase, but during the senescence of the culture. This also means that two-stage production must be developed; in stage one, growth occurs; in stage two, the product is produced. Frequently, the useful product is deposited in the vacuole, rather than being secreted to the medium. This means that the cells must be harvested and product extracted, adding expense and preventing continuous cultures from being developed. Finally, the high cost of the complex media required to maintain plant cells in culture adds to the difficulties of establishing a two-stage sterile process plant on a large scale.

The high-value products derived from plants are frequently plant secondary products – not produced as part of the major pathways of metabolism (*Table 14.2*). As such, they are complex compounds synthesized by multi-enzyme pathways that are often induced when a cell type is nearing the end of its life. Genetically engineering such pathways is difficult and initiating a pathway that is a characteristic of terminal differentiation in a cell culture is difficult. In many instances, secondary products are only produced in large quantities by highly specialized cell types; replicating the conditions that initiate the differentiation of such cells is also technically

Table 14.1. Key features of plant and microbial cells in culture

Feature	Micro-organism	Plant cell
Size	2 µm	>10 µm (and often colonies of cells)
Doubling time	Hours	Days
Length of culture cycle	Days	Weeks
Shear stress	Insensitive	Sensitive
Secreted product?	Yes – to medium	No – to vacuole
Production during growth?	Yes	No, often during senescence

Table 14.2. Some important plant secondary products

Family of compounds	Biosynthesis	Examples	Species
Terpenoids, isoprenoids	Mevalonic acid pathway, from acetyl CoA	Sterols and steroid derivatives	
		Cardiac glycosides like digoxin and digitoxin	Digitalis purpurea
		Plant hormones	
		Latex	Hevea brasiliensis
		Flavours like menthol	Mentha palustris
		Paclitaxel (anti-cancer agent)	Taxus brevifolia
Phenolics	Shikimic acid pathway	Tannins (flavours)	Numerous
		Flavonoids (pigments like anthocyanins)	
Alkaloids	Diverse origins	Caffeine, nicotine	Coffea arabica, Nicotiana tabacum
		Morphine, codeine	Papaver somniferum
		Atropine	Atropa belladonna
		Vinblastine, vincristine	Catharanthus roseus

very demanding. Cultured plant cells often produce low yields of secondary products in comparison with intact plants and this is directly related to the problems of obtaining sufficient differentiated material. There are, however, some examples of very high yields from cell cultures (*Table 14.3*; Misawa, 1994). Several of these, including the anti-microbial dyes shikonin and berberine and the medicinal compound ginsenoside, are currently industrially produced from plant cell culture and will be described in detail, as will work to produce the important anti-cancer agent paclitaxel.

3.   Apparatus for large-scale plant cultures

In most cases, plant suspension cultures are studied using cultures grown in conical flasks in an orbital incubator (Chapter 7). Sufficient aeration is maintained by the relatively high surface area to volume ratio of the liquid medium, and by agitation. Cells are also kept from aggregating or precipitating by agitation. Agitating and aerating a large volume of medium is much more difficult, especially as plant cells are easily damaged by mechanical forces and rapid pressure changes. Any apparatus also has to prevent contaminant organisms from entering.

Table 14.3. Some secondary products produced in high yield by plant cell cultures

Compound	Species	Yield (% dry weight)	
		Culture	Plant
Shikonin	*Lithospermum erythrorhizon*	20	1.5
Ginsenoside	*Panax ginseng*	27	4.5
Berberine	*Coptis japonica*	10	2–4
	or		
	Thalictrum minor		0.01
Rosmarinic acid	*Coleus blumei*	15	3
Anthraquinones	*Galium verum*	5.4	1.2
Nicotine	*Nicotiana tabacum*	3.4	2.0

From Misawara (1994).

Several designs of culture vessel have been devised for industrial-scale plant cultures (*Figure 14.1*). Vessels must have a means for agitation, which may be 'air-lift', that is, the effect of a rising column of air, a paddle or paddles near the bottom of the vessel, or rotation of the entire vessel. They will also have some method of temperature control, which may be a water jacket, and frequently have ports for adding and removing cells or media and for probes to monitor temperature and pH. Air inlets and other ports have a micro-pore filter system to maintain sterility. Industrial vessels for plant cell culture applications are referred to by some authors as fermentors. It is probably wise to differentiate the aerobic conditions required for plant cell cultures from those for a yeast fermentation, so the alternative names 'bioreactor' and 'culture vessel' are commonly used.

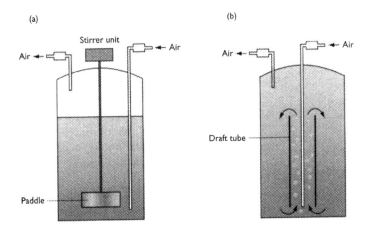

Figure 14.1

Two types of industrial-scale culture vessel. (a) Agitation is provided by an electrically driven paddle, while aeration is by a filtered air supply. (b) Both aeration and agitation are provided by the flow of air through a central glass pipe, called a draft tube, which causes a circulation of the medium as indicated by the arrows.

4.   Examples of industrial-scale production using cultured cells

4.1   Shikonin

Shikonin, a rich reddish-purple pigment that is also anti-microbial in activity, was first successfully produced by cell culture in 1974, by Tabata et al. in Japan. They initially grew callus of Lithospermum tinctoria (the natural source of shikonin) on Linsmaier and Skoog (LS) medium, with 10^{-6} M kinetin and 10^{-6} M 2,4-D for 3 months in the dark, then transferred it to fresh medium with IAA present. The system was developed by Fujita et al. (1985) into a highly productive two-stage culture process. Several factors were changed during development. The key components of the media were optimized to give two new media, MG-5 and MG-6. High-yielding cell lines were selected by obtaining protoplast by digesting the cell walls (Chapter 8) and then using a fluorescence-activated cell sorter (see Chapter 8) to select high-yielding individuals. In the process, called the Mitsui petrochemical process after the company that owns and operates it, cells are grown first on a multiplication medium, MG-5, and then transferred to a second medium, MG-6 to produce the dye in a second culture vessel. Best yields of dye are obtained when a high density of cells is achieved, with a final production rate of more than 4 g l^{-1} (Papageorgiou et al., 1999). The process is in routine commercial use.

4.2   Ginseng

Ginseng has been used since ancient times as a medicinal plant. Its roots contain a number of saponins and sapogenins, which are bio-active, including ginsenoside which is a sedative (Misawa, 1994). Demand for ginseng globally is high and the plant takes 4–7 years to accumulate the active ingredients in its roots. In Japan, the Nitto Denko Corporation is routinely producing $700 \text{ mg l}^{-1} \text{ day}^{-1}$ of ginsenoside from an industrial-scale suspension culture of Panax ginseng for pharmaceutical and nutritional use.

4.3   Paclitaxel

Paclitaxel is the generic name for the active ingredient of Taxol, a powerful anti-cancer agent produced by Bristol-Myers Squibb. Taxol is a diterpine amide that stabilizes microtubules and prevents their depolymerization, thereby preventing cell proliferation. It is used in the treatment of a range of cancers, including ovarian, breast and small-cell lung cancer.

Taxol was first isolated from the bark of the Pacific yew tree, Taxus brevifolia. Once approval had been gained for its clinical use, it was quickly evident that alternative sources were necessary, as treating one patient requires the bark of three 100-year-old trees. Organic synthesis of paclitaxel has not proved economic, but a semi-synthetic form has been derived from the needles and twigs of another member of the yew family, the Himalayan yew tree, Taxus bacatta, providing a renewable source. This form of paclitaxel has now replaced the drug derived from the bark of the Pacific yew; however, Bristol-Myers Squibb have worked with Phyton, a company specializing in plant tissue culture to develop a cell-culture technique. Phyton has collected and screened cell lines from every member of the Taxus family and used a combination of genetic and process modifications to maximize yield. The large-scale culture system is run according to current Good Manufacturing Practice standards and provides a stable supply of the product (Venkat, 1997). Other research groups are also actively developing cell cultures as a source of paclitaxel; for instance, Luo and co-workers in Wuhan, Peoples Republic of China, have developed a fed-batch system. Sucrose is fed to *Taxus chinensis* cells at $20\ g\ l^{-1}\ day^{-1}$ giving a maximum production of $19\ mg\ l^{-1}$ in 5-l culture vessels for 32 days (Luo *et al.*, 2002). With worldwide sales of around $1000 million per annum, securing a stable source of paclitaxel is a major research goal.

4.4 Hairy root cultures

When plant roots are transformed with *Agrobacterium rhizogenes*, they can be subcultured and grown indefinitely on media without hormones (Chapter 10). The roots display proliferation of lateral roots and can show very rapid growth, with a doubling time of between a day and a week (Bourgaud *et al.*, 2001). The great advantage of hairy roots is that they produce secondary metabolites as they grow; thus, systems where biomass increase is optimized also give the highest yields of product. Hairy roots can be grown in large-scale liquid cultures and are generally considered to be less susceptible to mechanical damage than cell suspensions are (Bourgaud *et al.*, 2001).

In general, the metabolism of hairy roots is similar to that of intact roots, though levels of secondary products may differ. Their use has been investigated for production of tropane alkaloids, betacyanin, nicotine, scopolamine serotonin and other compounds of medicinal relevance (Misawa, 1994). There are also a variety of applications for genetically modified hairy roots, for instance in the production of monoclonal antibodies (Wongsamuth and Doran, 1997).

4.5 Immobilized cell cultures

Immobilized cell cultures have many potential advantages over liquid suspensions; however, to be effective, it is essential that the product be secreted by the cells into the medium. When this is the case, medium flowing over the cells can be taken and extracted, while the cells remain alive to generate more product. Unfortunately, most plant cells in culture generate products that are stored in the vacuole and immobilized plant cells have not been commercialized.

Several methods exist for immobilization. Early studies embedded the cells in a gelatinous matrix like agar, agarose, carageenan and alginate. Absorption of cells to the surface of polyurethane foam has also been successful. *Capsicum frutescens* cells produced 50-fold the yield of capsaican as suspension cultures and *Datura innoxia* produced comparable yields of tropane alkaloids to the parent plant when production by suspension cultures was much less (Misawa, 1994). It is clear that immobilized cells will have many advantages if cell systems can be engineered in which products are secreted.

4.6 Biotransformations (*Table 14.4*)

Cells in suspension or immobilized cells may be used to modify a chemical to yield a high-value product. A precursor of the final product is added either to the suspension culture medium or in continuous flow over immobilized

Table 14.4. Examples of biotransformations by plant tissue cultures

Species	Feedstock	Product	Reference:
Digitalis lanata	β-methyldigitoxin	β-methyldigoxin	Spieler et al. (1985)
Catharanthus roseus	Catharanthine and vindoline	Vinblastine	Endo et al. (1987)

cells. Spieler *et al.* (1985) achieved the production of β-methyldigoxin by *Digitalis lanata* with β-methyldigitoxin as a feedstock. Vinblastine, a commonly used anti-cancer drug, is an indole alkaloid derived from the *Vinca* family. Vinblastine is formed from two alkaloids, catharanthine and vindoline, and is extracted from *Catharanthus roseus* plants. Endo *et al.* (1987) showed the potential for biotransformation by supplying the two substrates, catharanthine and vindoline, to *C. roseus* suspension cultures, which then yielded vinblastine. A more successful approach has been to use the *C. roseus* cultures to synthesize catharanthine using a supplemented MS medium, then to extract the catharanthine and react it in the presence of an iron catalyst with vindoline. This combined process has been developed by the Mitsui Petrochemical Company in Japan (Misawa, 1994).

References

Bourgaud, F., Gravot, A., Milesi, S. and Gontier, E. (2001) Production of plant secondary metabolites: a historical perspective. *Plant Sci.* **161**: 839–851.

Endo, T., Goodbody, A., Vukovic, J. and Misawa, M. (1987) Biotransformation of anhydrovinblastine to vinblastine by a cell-free-extract of *Catharanthus roseus* cell-suspension cultures. *Phytochemistry* **26**: 3233-3234.

Fujita, Y., Takahashi, S. and Yamada, Y. (1985) Selection of cell-lines with high productivity of shikonin derivatives by protoplast culture of *Lithospermum-Erythrorhizon* cells. *Agric. Biol. Chem.* **49**: 1755.

Luo, J., Mei, X.G., Liu, L. and Hu, D.W. (2002) Improved paclitaxel production by fed-batch suspension cultures of *Taxus chinensis* in bioreactors. *Biotechnol. Lett.* **24**: 561–565.

Misawa, M. (1994) *Plant Tissue Culture: An Alternative for Production of Useful Metabolite.* FAO Agricultural Services Bulletin 108. FAO, Rome.

Papageorgiou, V.P., Assimopoulou, A.N., Couladourous, E.A., Hepworth, D. and Nicolaou, K.C. (1999) The chemistry and biology of alkannin, shikonin and related naphthazarin natural products. *Angew. Chem.* **38**: 270–300.

Spieler, H., Alfermann, A.W. and Reinhard, E. (1985) Biotransformation of beta-methyldigitoxin by cell-cultures of *Digitalis lanata* in airlift and stirred tank reactors. *Appl. Microbiol. Biotechnol.* **23**: 1–4.

Tabata, M., Mizukami, H., Hiraoka, N. and Konishima, M. (1974) Pigment formation in callus cultures of *Lithospermum erythrorhizon*. *Phytochemistry* **13**: 927–932.

Venkat, K. (1997) Paclitaxel production through plant cell culture: an exciting approach to harnessing biodiversity. IUPAC, http://www.iupac.org/symposia/phuket97/venkat.html

Wongsamuth, R. and Doran, P.M. (1997) Production of monoclonal antibodies by tobacco hairy roots. *Biotechnol. Bioeng.* **54**: 401–415.

Chapter 15

Prospects and future challenges

1. Recent developments

The rapid development of techniques for plant cell and tissue culture over the last three decades has had a major impact on many areas of plant science. During this period, the number of technical reports detailing protocols for the successful application of tissue culture technology to a wide range of species has dramatically increased. It is now possible to regenerate plants from a range of tissue explants from many plant species, including most of the important crop, fruit and ornamental species. More recently, developments in the science of plant cell and tissue culture have combined with rapid advances in molecular biology to drive a new revolution in plant biology and to launch the field of plant biotechnology. Some remarkable breakthroughs have been achieved and many procedures have been devised that are currently valuable tools for basic research or have been successfully exploited in commercial applications.

2. The future

It can be anticipated that plant biotechnology will continue to evolve in response to the immense challenge that is presented by the vast range of biological diversity in plant species. The advancement of established areas and the exploration of new directions will require the constant development and refinement of cell culture technology. New possibilities making use of the information obtained from genome sequencing of crops like rice, the application of technologies such as genomics, proteomics and metabolomics are being explored, and there exists the likelihood of impressive breakthroughs in the foreseeable future.

2.1 Improved regeneration

The combined techniques of tissue culture and molecular biology have advanced to the point that transgenic plants can be created by the introduction of genes from almost any source. Furthermore, the exploitation of these technologies has led to the commercialization of transgenic crops. However, as a general strategy, efficient transformation and regeneration remains a feat of skill because each species or genotype requires unique culture conditions. The manipulation of species that have not been previously transformed or cultured requires either the careful adaptation of established protocols or the precise development of new methods, taking into account factors such as the tissue, media composition and culture conditions. Central to this optimization of culture conditions is the achievement of stable and efficient transgene integration and expression. Sometimes these approaches are insufficient to provide reliable protocols and alternative strategies based on the application of genomic methods to genetic mapping are being investigated.

Genetic mapping: quantitative trait loci (QTL)

In plants, agronomic traits such as yield and disease resistance are complex because they are often controlled by several genes (i.e. polygenic), each segregating according to Mendelian laws. In addition, these traits are also influenced by environmental factors. These traits are referred to as quantitative traits and the loci controlling these traits are known as quantitative trait loci.

Considering their importance, the improvement of quantitative traits is among the primary objectives of many plant-breeding programmes. In order to identify generation lines that contain the QTL alleles that contribute to the trait under selection, molecular markers associated with individual characters are identified in segregating populations and used to construct linkage maps that dissect the quantitative trait into single Mendelian components. This linkage data is used to estimate the number of loci controlling the genetic variation, to map their position in the genome and to elucidate the environmental influence on gene action.

Various types of markers are used to map QTLs. Originally isoenzymes were used as molecular markers, but recently more precise molecular DNA markers such as RFLP, RAPD, AFLP and sequence repeats (SR) of DNA sequences including minisatellites (10–45 bp) and microsatellites (2–6 bp). RAPD, AFLP and SR techniques are based on PCR and are quicker than RFLP techniques. Once the associations of markers with QTLs have been defined, phenotypes can be predicted based on the absence or presence of markers. In addition, mapping techniques can be used to identify genes for a particular character, which can be cloned, sequenced and transferred to other species.

Since the inheritance of the capacities for callus formation and for plant regeneration is complex, QTL for these characters that are constant over a range of culture conditions can be used to screen and select suitable germplasm for transformation, culture and regeneration. This technique is known as marker-assisted selection (MAS). In some crops, for example maize and rice, anther culturability is a quantitative trait controlled by nuclear encoded genes (Beaumont et al., 1995; Kwon et al., 2002). QTL information for this trait should help in selecting genotypes that are culture responsive and to transfer the regenerability trait to genotypes that respond poorly. In rice, this would aid haploid breeding in indica cultivars that display low plant regeneration from anther culture.

2.2 Production of pharmaceutical proteins in plant cell and organ cultures

The idea that plants could be genetically engineered to produce recombinant proteins on a commercial scale for animal and human diagnostics and therapeutics is not new. Hiatt et al. (1989) were the first to demonstrate the expression of functional antibodies in tobacco leaves. Since then the technology for the production of biopharmaceuticals in plants has progressed and it is now recognized that plant systems are a cost-effective alternative to microbial and animal cell cultures for the production of proteins such as enzymes, antibodies and vaccines. Currently, there is a high demand for cheap sources of human therapeutic proteins like erythropoietin (used to treat anaemia) and insulin (used to treat diabetes). In the future, this demand is likely to be extended to a new generation of protein pharmaceuticals that will be developed by exploiting information obtained from the human genome sequence. Plants are potentially a cheap source of recombinant pharmaceuticals, and several biotechnology companies are now patenting plant expression systems for drug production and actively conducting field trials. Moreover, a few of the plant-derived products have already made it into the early stages of human clinical trials. A number of plant expression systems are being developed for the production of biopharmaceuticals.

There are several economic and technical advantages that favour the use of plant systems as potential large-scale sources of these recombinant products:

- low risk of product contamination by mammalian-specific pathogens;
- correct polypeptide folding and multimeric assembly;
- the relative ease of introducing new transgenes by sexual crossing;
- avoidance of ethical problems associated with the use of animals.

Plant tissue culture production or whole plant agriculture?

Like whole plants, plant cell and organ cultures such as hairy roots that have been transformed with mammalian genes are capable of producing a wide range of recombinant proteins. Three main methods for the commercial production of recombinant proteins in plants are being explored:

- large-scale agricultural production by field-grown crops;
- agricultural production using greenhouse-cultivated plants;
- plant cell and organ cultures grown in bioreactors.

Several factors need to be considered in deciding which plant system is appropriate for production. In economic terms, field-grown production is by far the most cost effective. In some cases where the product will be delivered in the edible parts of plants (e.g. edible vaccines) without the need for extraction and purification, the downstream processing costs are greatly reduced. However, environmental concerns and regulatory issues arising from the open-field cultivation of transgenic crops could be formidable obstacles to this method of production.

Some of the concerns like environmental contamination can be met through rigorous containment by using greenhouse cultivation. Nonetheless, recombinant protein production in greenhouse-cultivated plants is more expensive than in field-grown crops by an order of magnitude.

At present, although plant cell and organ cultures are unlikely to be cost competitive with field-grown and greenhouse-grown plants for the large-scale production of foreign proteins, cell and organ cultures have a number of inherent technical advantages that could be beneficial for production. For example, cell and tissue cultures grown in bioreactors are more amenable to manipulation of growth and protein production than cultivated plants. Development of culture regimes to significantly improve protein expression and accumulation would therefore increase the economic viability of plant cell cultures. Another advantage of cell cultures is that their time-scale for growth and production is much shorter than that for whole plants, days or weeks compared with months. This means that supply can more easily be matched to clinical demands. In addition, the environmental risks associated with transgenic plants are avoided by growing the transformed plant cells in bioreactors. Product purification and processing is greatly facilitated in plant cell culture expression systems where the product is secreted to the medium. Protein can be separated from the cells or tissue by a non-destructive step.

Based on these considerations, it has been suggested that plant cell cultures grown in bioreactors may fill a production niche for high-value, high-purity, speciality therapeutic proteins (Doran, 2000).

The targeting problem

It has already been noted that many high-value plant products are deposited in the vacuole after synthesis and are not secreted to the medium. This has a number of implications:

- continuous culture from which product is harvested from the medium is not possible;
- cells must be lysed and extracted after harvest;
- product may be degraded by the acidic contents of the vacuole or degraded during extraction by mixing with enzymes of the cell contents.

Engineering cells in which products are secreted to the medium would therefore greatly enhance the usefulness of cell plant cultures. To date, such a fundamental understanding of the plant secretory pathway is lacking. While great advances are being made in understanding traffic and targeting of products, development of cell lines with altered secretion should be a major research goal. Generating inducible cell lines would be even more valuable as secretion to the medium could be initiated only when the product was being generated in quantity.

The sugar problem

Many recombinant human therapeutic proteins like antibodies and blood proteins have been successfully expressed in transgenic plants. In most cases, the plant-produced proteins are identical to those formed in mammalian cells with respect to amino acid sequence, conformation and biological activity. However, these proteins are N-glycosylated glycoproteins and the final steps of N-glycan processing in plants and mammals differ, resulting in the formation of proteins with organism-specific glycans. Although these differences may not affect the activity of the protein, other clinically important properties such as antigenicity and rate of blood clearance may be significantly affected. In a recent comparative study of the N-glycosylation of a monoclonal antibody (Guy's 13 IgG1) expressed in mouse and transgenic tobacco (Cabanes-Macheteau et al., 1999), although core glycans were similar, the plant protein had complex N-glycans with (1,2)-xylose and (1,3)-fucose residues. The plant also lacked terminal sialylation. The plant specific (1,2)-xylose and (1,3)-fucose residues are highly immunogenic in mammals and are key epitopes responsible for allergenic responses in humans. Lack of terminal sialylation is known to reduce the *in vivo* half-life of the proteins.

Strategies to avoid these plant glycosylation problems include producing transgenic plants lacking specific enzymes in the glycosylation pathway or alternatively modifying the protein after extraction.

The problem of multiple gene expression

Most of the high-value products obtained from plants are the result of complex metabolic pathways and are often at the termini of such pathways (Chapter 3). Metabolic control analysis has also revealed that achieving significant alteration in the activity of metabolism requires more than the alteration in activity of single genes. To date, most successful genetic modifications have essentially involved only single gene changes. Modification of the activity of several genes, or introducing a series of genes required, for instance for synthesis of a secondary product would represent a major advance and open up numerous possibilities both for factory-scale *in vitro* systems, where a pathway could be engineered into a cell type which is amenable to culture, and for field production. Apart from the experimental difficulty of multiple transformation, such a multi-gene system would require appropriate co-regulated promoters for each gene to ensure appropriate levels of activity.

2.3 Future prospects

Colinearity and synteny

Information from the completed genome sequence of two model species, *Arabidopsis* for dicots and *Oryza* (rice) for monocots, together with the increasing availability of expressed sequence tags (ESTs) for other species, will make it easier by comparative genomics to locate genes in other species. Genes with related function are grouped together in related species (properties described as colinearity and synteny). Thus, the fact that genes are located together in one species can be used to locate similar gene groupings in another species (Flavell, 1995). For instance, colinearity of syntenic loci is well conserved in the family *Poacea*. Genes identified and located in rice can then be used to identify and locate homologous genes in distant relatives of the family like sugarcane.

The future development of these strategies to harness traits like *in vitro* culturability or to exploit plant genes for molecular farming will make plant tissue culture and associated techniques more effective research tools as well as widen their applicability.

Bioinformatics

The combined fields of genomics, proteomics and metabolomics are rapidly yielding large amounts of information about the genes, proteins and pathways of plant productivity. Understanding these elements permits a new level of rational design in the modification of cell culture for the production of high-value products. The combination of genome data with germplasm data will also advance plant breeding, both of conventional and genetically modified crops. Interpreting the volumes of information generated by these approaches will require both computational and interpretive expertise, and the role of the bioinformatician will become increasingly important.

2.4 Concluding remarks

Plant cell and tissue culture is a fundamental tool in pure and applied plant science. The last four decades have seen rapid and exciting advances equal to any seen in biology and the applications of the technology have had global implications. In some areas, the application of transgenic technology has met with consumer resistance and environmental caution; techniques viewed with concern in some nations are finding rapid acceptance in others. In other areas, like in the use of cultures in bioreactors, apart from in a few instances (Chapter 14), difficulties and costs of application have restricted their use. There is no doubt, however, that the combination of genetic modification and cell and tissue culture applied with due caution presents immense opportunity for progress. Such progress can be expected in the improvement of production by factory-scale contained systems and in the field, both with genetically modified and non-genetically modified, clonally propagated plants. At the heart of such advance must be an improved understanding of plant biology – an understanding in part to be gained by the application of plant cell and tissue culture itself.

References

Beaumont, V.H., Rocheford, T. and Widholm, J.N. (1995) Mapping the anther culture response genes in maize (*Zea mays* L.). *Genome* 38: 968–975.

Cabanes-Macheteau, M., Fitchette-Laine, A.C., Loutelier-Bourhis, C., Lange, C., Vine, N.D., Ma, J.K.C., Lerouge, P. and Faye, L. (1999) N-glycosylation of a mouse IgG expressed in transgenic tobacco plants. *Glycobiology* 9: 365–372.

Doran, P.M. (2000) Foreign protein production in plant tissue cultures. *Curr. Opin. Biotechnol.* 11: 199–204.

Flavell, R.B. (1995) Plant biotechnology R & D – the next ten years. *Trends Biotechnol.* 13: 313–319.

Hiatt, A.C., Cafferkey, R. and Bowdish, K. (1989). Production of antibodies in transgenic plants. *Nature* 342: 76–78.

Kwon, Y.S., Kim, K.M., Eun, M.Y. and Sohn, J.K. (2002) QTL mapping and associated marker selection for the efficacy of green plant regeneration in anther culture rice. *Plant Breeding* 121: 10–16.

Suppliers of chemicals, apparatus and cell culture products

Bambelt Instruments BV
J. F. Vlekkestraat 272A
4705 AJ Roosendaal
The Netherlands
http://www.bambelt.com/

Bellco Glass, Inc.
340 Edrudo Road
P.O. Box B
Vineland
NJ 08360
USA
http://www.bellcoglass.com/

Biolink New Zealand Limited
P.O. Box 65–281
Mairangi Bay Auckland 1330
New Zealand
http://www.biolink.co.nz/

Biomark Laboratories
Apollo Green, 441/3 Somwar Peth
Near Meherbaba Centre
Pune 411 011
India
http://www.biomarklabs.com/plant.html

Carolina Biological Supply Company
2700 York Road
Burlington, NC 27215
USA
http://www.carolina.com/

Dent-Eq
No. 101/1, Surveyor St
Basavanagudi
Bangalore 560004
Karnataka
India
http://www.dent-eq.com/

Fisher Scientific Worldwide
Export Division
50 Fadem Road
Springfield
NJ 07081-3193
USA
https://www3.fishersci.com

Gibco Laboratories
3175 Staley Road
Grand Island
NY 14072
USA
http://www.lifetech.com

ICN Biomedicals, Inc.
3300 Hyland Ave
Costa Mesa
CA 92626
USA
http://www.icnbiomed.com/

Kitchen Culture Kits, Inc.
905 Champions Drive
Lufkin
TX 75901-7235
USA
http://www.kitchenculturekit.com/

Lehle Seeds
1102 South Industrial Blvd
Suite D
Round Rock
TX 78681
USA
http://www.arabidopsis.com/

Melford Laboratories Ltd
Bildeston Road
Chelsworth
Ipswich
IP7 7LE
Suffolk
www.melford.co.uk

Merck & Co., Inc.
Kelco Division
Rahway
New Jersey
USA
www.serva.de/products/knowledge/081029.shtml

Osmotec
Distributed worldwide by Sigma
http://www.osmotek.com/

PhytoTechnology Laboratories
P.O. Box 13481
Shawnee Mission
KS 66282
USA
http://www.phytotechlab.com/

Plant tissue culture network
An online directory of suppliers
http://aggie-horticulture.tamu.edu/tisscult/database/index.html

Research Products International Corp.
410 N. Business Center Drive
Mount Prospect
IL 60056-2190
USA
http://www.rpicorp.com/index.php

Sigma Chemical Company
P.O. Box 14508
St. Louis
MO 633178
USA
http://www.sigmaaldrich.com/

Tissue Culture Network Supplier Database
http://horticulture.tamu.edu:7998/tissueculturesuppliers/index.html

VWR Scientific
1310 Goshen Parkway
West Chester
PA 19380
USA
Orders: 1-800-932-5000
http://www.vwr.com/

 **The Basics**

Glossary

Abiotic	Not of biological origin
Abscission	Separation of a plant leaf or fruit from the plant body, often as a result of a programmed pathway of development
Agrobacterium rhizogenes	A soil bacterium related to *A. tumefaciens* that transforms plant roots inducing root hair proliferation forming 'hairy roots'
Agrobacterium tumefaciens	A soil bacterium capable of infecting and transforming plant material to form the crown gall tumour and from which the modified plasmids used in plant transformation are derived. Non-violent strains are used in plant genetic engineering
Androgenesis	Creating a progeny by doubling the male (haploid) genome
Angiosperm	Flowering plants that reproduce by seeds covered in a fruit
Aneuploidy	Having more or less than an integral multiple of the haploid number of chromosomes
Apomixis	Of asexual origin, e.g. derived from unfertilized eggs or ovule tissue
Aseptic	In the absence of contaminating bacteria, fungi, spores, viruses, etc.
Autotrophic	Capable of generating its own food supply (e.g. by photosynthesis)
Auxin	Generic term for a family of plant growth substances (hormones) involved in the regulation of growth and development
Bioreactor	Large-scale culture vessel used for plant (or other) cell and tissue culture
Callus	Amorphous mass of cells generated either as a wound response or in tissue culture
Chimaera	Organism or genetic construct containing genetic material originating from two or more different organisms
Chlorenchyma	Parenchyma-containing chloroplasts
Clonal propagation	Production of genetically identical progeny by asexual means
Cloning vector	An agent used in gene cloning to insert a foreign DNA fragment into the genome of a host cell

Coconut water (milk)	The liquid endosperm of the coconut, used as a medium additive in some plant cell and tissue culture techniques
Codon	Sequence of three adjacent nucleotides (bases) in DNA or RNA encoding an amino acid
Collenchyma	Supportive tissue with thickened cell walls
Cotyledons	Leaves formed as part of the embryo
Cultivar	A 'cultivated variety' comprising a group within a species that when reproduced together sexually, or asexually, retains its characteristics
Cybrid	Cytoplasmic hybrid in which a cytoplasmic sterile line is fused by cell fusion with another cell line, mixing the cytoplasmic genomes
Cytokinin	Family of growth substances (hormones) involved in regulating growth and development
Cytoplasmic male sterility	Failure to produce viable pollen
Differentiation	Process in development whereby cells and tissues become specialized to a particular structure or function
Diploid	Having twice the haploid chromosome number
Embryogenesis	Formation of embryos
Endosperm	Storage tissue of cereal grains
Epidermis	Outer, protective cell layer of organs
Epigenetic	Stably transmitted modifications that result in stable phenotype differences even when the parental cells are genotypically identical
Genetic modification (GM)	Insertion of a foreign gene or promoter, or other modification of a genome in order to obtain a transgenic organism
Genome	The chromosomal genetic material of the cell. Plant cells may have three genomes, nuclear, mitochondrial and plastid
Genomics	The study of the physical organization and function of the genome
Germplasm	Generic term for the genetic information contained within living material, either as seed, plants or tissue or cell culture
Glyphosate	A herbicide, Trade names include Gallup, Landmaster, Pondmaster, Ranger, Roundup, Rodeo, and Touchdown
Gymnosperm	Second division of the Spermatophyta (cf. Angiospermae), of which the conifers are a characteristic example
Haploid	Having a single set of unpaired chromosomes.
Haemocytometer	Device used to count cells
Hairy roots	Roots transformed with *A. rhizogenes* with a 'hairy' appearance due to root hair formation
Heterokaryons	Cells with more than one nucleus

Heterologous	Involving material from two or more origins (e.g. heterologous recombination – the combination of genetic material from two parental lines)
Heterotrophic	Requiring an external food source (cf. autotrophic)
Homozygous	Organism or cell where the alleles at a given locus on homologous chromosomes are identical
Insert	DNA inserted into a vector and ultimately a genome used in transformation
Hypocotyl	Seedling stem below cotyledons
Laminar flow cabinet	Apparatus used to create an aseptic work area in a tissue culture laboratory using a filtered airflow
Ligation	Joining genetic material together to form an artificial construct
Meristem	Zone of cell division in the plant
Metabolomics	Study of cells at different physiological states by the analysis of the entire complement of small molecular weight metabolites
Micropropagation	Generation of clonal progeny by asexual means using tissue culture
Microsatellite DNA	Short repeated pieces of DNA (2–6 bp)
Minisatellite DNA	Short repeated pieces of DNA (10–45 bp)
Mitotic index	The proportion of cells at any given stage of mitosis
Mixoploidy	The presence of cell lines with a different genetic constitution in an individual
Mutagenesis	Induced mutation of the genome, either natural or artificial
Opines	Nitrogen containing compounds produced after *Agrobacterium* infection
Osmostabilizer	Compound used to maintain the osmotic potential of a solution
Ovule	Structure in gymnosperms and angiosperms containing the egg cell, which after fertilization will develop into the seed
Parenchyma	Tissue made up of thin walled cells
Parthenocarpic	Formation of fruit without fertilization
Periclinal	Division perpendicular to the plane of a cell sheet
Phloem	Vascular conductive tissue primarily responsible for the transport of carbohydrates and organic compounds in the plant
Plasmid	A small, circular double-stranded DNA molecule, existing inside a host such as a bacterium, which replicates independently of the host genetic material and can be engineered to contain genes of interest
Plasmodesmata	Intercellular connections between plant cells
Plasmolysis	Osmotically induced shrinkage of the cell contents (protoplast) from the cell wall

Polygenic	Controlled by several genes
Polyploidy	Having three or more times the haploid (n) number of chromosomes
Primer	A sequence of nucleotides used to bind to a DNA template to initiate copying of the DNA sequence
Proteomics	Quantification and description of the protein composition of a cell or organism
Protoderm	The primary meristem that gives rise to the epidermis
Protoplasts	Cells from which the cell wall has been removed
Quantitative trait loci (QTL)	Polymorphic loci that contain alleles that differentially affect the expression of a phenotypic trait and used as a marker for variation
Random amplification polymorphic DNA (RAPD)	A technique to compare DNA sequence by amplifying fragments of DNA by PCR using short primers
Restriction fragment length polymorphism (RFLP)	A DNA digestion technique used to detect differences in gene sequence between related plants
Sclereids/ Sclerenchyma	Supportive cells and tissue within the plant
Senescence	A genetically programmed stage of development in which tissues age, die and may abscind
Somaclonal variation	Genotypic variation from a variety of causes resulting in variation in the progeny of clonal propagation
Somatic embryos	Embryos produced from somatic cells (i.e. in tissue culture); cf. zygotic embryos produced by normal plant reproduction
Strobilus	Cones – the fruiting body of conifers
Synteny	Genes arranged in the same order on chromosomes of different species
Totipotency	The ability of a differentiated cell to retain and express the entire genome required to produce an entire organism
Transgenic	Containing a foreign gene or genes
Transposon	A segment of DNA that is capable of independently replicating itself and inserting the copy into a new position within the same or another chromosome or plasmid
Triploid	A plant or cell containing three times the haploid number of chromosomes
Vector	See cloning vector
Xylem	Vascular conductive tissue primarily responsible for the transport of water and inorganic compounds in the plant

Index

Milton Keynes UK
Ingram Content Group UK Ltd.
UKHW051924141024
449569UK00027B/1349